# Lecture Notes in Computer Science     13825

More information about this series at https://link.springer.com/bookseries/558

Xianfeng Zhao · Zhenjun Tang ·
Pedro Comesaña-Alfaro · Alessandro Piva (Eds.)

# Digital Forensics and Watermarking

21st International Workshop, IWDW 2022
Guilin, China, November 18–19, 2022
Revised Selected Papers

*Editors*
Xianfeng Zhao (iD)
Chinese Academy of Sciences
Institute of Information Engineering
Beijing, China

Pedro Comesaña-Alfaro (iD)
Universidade de Vigo
Vigo, Spain

Zhenjun Tang (iD)
Guangxi Normal University
Guilin, China

Alessandro Piva (iD)
University of Florence
Florence, Italy

ISSN 0302-9743          ISSN 1611-3349 (electronic)
Lecture Notes in Computer Science
ISBN 978-3-031-25114-6          ISBN 978-3-031-25115-3 (eBook)
https://doi.org/10.1007/978-3-031-25115-3

This Springer imprint is published by the registered company Springer Nature Switzerland AG
The registered company address is: Gewerbestrasse 11, 6330 Cham, Switzerland

# Preface

The International Workshop on Digital-Forensics and Watermarking (IWDW) is a premier forum for researchers and practitioners working on novel research, development and application of digital watermarking, data hiding, and forensic techniques for multimedia security. The 21st International Workshop on Digital-Forensics and Watermarking (IWDW 2022) was held in Guilin, China, during November 18–19, 2022, organized by Guangxi Normal University and sponsored by the Institute of Information Engineering, Chinese Academy of Sciences. As many as 226 people attended the workshop physically and 294 people attended it online.

IWDW 2022 received 30 submissions. Each of them was assigned to at least two members of the Technical Program Committee for review. The decisions were made on a highly competitive basis. Only 14 submissions were accepted according to the average score ranking. The accepted papers cover important topics in current research on multimedia security, and the presentations were organized into three sessions including "Steganology", "Forensics and Security Analysis", and "Watermarking". There were two invited keynotes: "Towards Better Generalization Capability of Face Spoofing Detection" by Haoliang Li from City University of Hong Kong, and "Recent Advances in Stegomalware: Development Trends and Detection Opportunities" by Wojciech Mazurczyk from Warsaw University of Technology and FernUniversität Hagen. In addition, IWDW 2022 kicked off the International Comparative Evaluation of STC (Syndrome Trellis Code) and SPSC (Sub-Polarized Steganographic Code) to promote research on steganographic codes.

We would like to thank all of the authors, committee members, reviewers, keynote speakers, session chairs, volunteers, and attendees, who once again made a memorable IWDW. And we appreciate the generous support from the organizer and sponsors. Finally, we hope that readers will enjoy this volume and find it rewarding in providing inspiration and possibilities for future work.

December 2022

Xianfeng Zhao
Zhenjun Tang
Pedro Comesaña-Alfaro
Alessandro Piva

# Organization

## General Chairs

Yun-Qing Shi           New Jersey Institute of Technology, USA
Shichao Zhang          Guangxi Normal University, China
Xianxian Li            Guangxi Normal University, China

## Technical Program Chairs

Xianfeng Zhao          Chinese Academy of Sciences, China
Zhenjun Tang           Guangxi Normal University, China
Pedro Comesaña-Alfaro  University of Vigo, Spain
Alessandro Piva        University of Florence, Italy

## Program Committee

Patrick Bas            University of Lille, France
Roberto Caldelli       University of Florence, Italy
François Cayre         Grenoble-INP GIPSA-Lab, France
Marc Chaumont          University of Montpellier, France
Rémi Cogranne          University of Technology of Troyes, France
Dinu Coltuc            Valahia University of Targoviste, Romania
Isao Echizen           National Institute of Informatics, Japan
Yongjian Hu            South China University of Technology, China
Guang Hua              Wuhan University, China
Jiwu Huang             Shenzhen University, China
Yongfeng Huang         Tsinghua University, China
Xiangui Kang           Sun Yat-Sen University, China
Sang-ug Kang           Sangmyung University, Korea
Stefan Katzenbeisser   University of Passau, Germany
Christian Kraetzer     Otto-von-Guericke University Magdeburg,
                       Germany
Minoru Kuribayashi     Okayama University, Japan
Xiaolong Li            Beijing Jiaotong University, China
Bin Li                 Shenzhen University, China
Qingzhong Liu          Sam Houston State University, USA
Wojciech Mazurczyk     Warsaw University of Technology, Poland
Akira Nishimura        Tokyo University of Information Sciences, Japan

| | |
|---|---|
| Fernando Pérez-González | University of Vigo, Spain |
| Tomas Pevny | University of Prague, Czech Republic |
| Andreas Uhl | University of Salzburg, Austria |
| Luisa Verdoliva | University of Naples Federico II, Italy |
| Kai Wang | University of Grenoble Alpes, France |
| Hongxia Wang | Sichuan University, China |
| Koksheik Wong | Monash University, Malaysia |
| Weiqi Yan | Auckland University of Technology, New Zealand |
| Weiming Zhang | University of Science and Technology of China, China |
| Xinpeng Zhang | Fudan University, China |
| Yao Zhao | Beijing Jiaotong University, China |

## Organizing Chair

| | |
|---|---|
| Jinyan Wang | Guangxi Normal University, China |

## Organizing Committee

| | |
|---|---|
| Hanyun Zhang | Guangxi Normal University, China |

## Additional Reviewers

Yun Cao
Peisong He
Lincong Li
Zehua Ma
Yanzhen Ren
Ante Su
Yuyang Sun
Chunpeng Wang

Li Xiang
Mengyao Xiao
Bowen Yang
Qiyi Yao
Xiaowei Yi
Xinquan Yu
Chunqiang Yu

# Sponsors

Chinese Academy of Sciences

Guangxi Normal University

New Jersey Institute of Technology

Springer

# Contents

# Steganology

# High-Performance Steganographic Coding Based on Sub-Polarized Channel

Haocheng Fu[1,2], Xianfeng Zhao[1,2], and Xiaolei He[1,2(✉)]

[1] State Key Laboratory of Information Security, Institute of Information Engineering, Chinese Academy of Sciences, Beijing 100195, China
{fuhaocheng,zhaoxianfeng,hexiaolei}@iie.ac.cn
[2] School of Cyber Security, University of Chinese Academy of Sciences, Beijing 100195, China

**Abstract.** Steganographic coding is the core problem of modern steganography under the minimizing distortion model. Syndrome-Trellis Codes, designed with the Viterbi algorithm of convolutional codes, have been the only coding scheme approaching the rate-distortion bound for almost a decade. Although polar codes has shown to have the potential to construct optimal coding schemes, the low diversity of steganographic coding negatively affects the security of steganography, and the performance of the state-of-the-art coding schemes is still unsatisfactory for large-scale applications. This paper proposes a high-performance steganographic coding scheme, named Sub-Polarized Steganographic Codes (SPSC), with near-optimal efficiency and lower computational complexity. By optimal embedding theory and channel polarization, we first establish a universal model of polarized steganographic channels for the coding of steganography. Based on this, the steganographic channels are divided into a combination of multiple sub-polarized channels to construct efficient steganographic codes by updating sequential bit-wise coding to segmented coding. The proposed coding scheme is also evaluated under four polarized channels with typical patterns, of which corresponding sub-coding schemes are also illustrated. Experimental results show that the proposed steganographic coding scheme improves security by increasing the embedding efficiency of steganography, and significantly decreases the computational complexity.

**Keywords:** Covert communication · Steganographic coding · Sub-channel polarization · Successive cancellation

## 1 Introduction

Steganography is an important branch of information hiding. By embedding secret messages into naturally-looking covers, it establishes covert communication while not arousing the attention of others [11]. Modern steganography

This work was supported by NSFC under 61972390, 61902391, 61872356 and 62272456, and National Key Technology Research and Development Program under 2022QY0101.

prefers using digital media to perform steganographic embedding for its wide-spreading on the Internet. Based on defined distortion of modification of each cover element, the most modern schemes achieve satisfactory undetectability by minimizing total embedding distortion under Payload-Limited Sender (PLS) and Distortion-Limited Sender (DLS) problems. For digital images, several heuristic-designed cost functions, such as HUGO [25], HILL [22], UNIWARD [20], UED [15] and UERD [16], were proposed for cover in both spatial and JPEG domain, which achieve high security performance even the distortion is considered additive.

To minimize the total distortion, Crandall [5] first proposed the concept of matrix embedding by introducing the parity-check matrix of the error-correction codes. Based on this, multiple linear coding schemes, such as Hamming [30], BCH [31,32] and LDPC [6,12] were utilized to optimize with the constant distortion profile globally. Thereafter the wet paper codes (WPC) [13,14] were proposed to for distortion consisting of modifiable (dry) and unmodifiable (wet) elements since they believe the modification tends to be performed on more secure areas.

In advanced steganographic schemes, coding schemes are considered to be adopted continuously instead of multi-level discrete distortion optimization. Therefore, Filler *et al.* [9,10] proposed the first near-optimal steganographic codes, named Syndrome-Trellis Codes (STC), by the Viterbi decoding algorithm for arbitrary distortion. After that, no more practical steganographic coding scheme has been proposed for almost a decade, which might bring negative impact on the security of steganography. Recently, since polar codes [2] is proved to be able to achieve channel capacity for any symmetric binary input discrete memoryless channels (B-DMC), in [8], Diouf *et al.* constructed the first coding scheme of steganography with the Successive Cancellation (SC) decoding algorithm of polar codes, which revealed that steganographic codes based on polar codes have the ability to reach the rate-distortion bound of content-adaptive steganography. Thereafter Li *et al.* [23] designed another near-optimal coding scheme based on polar codes for cover with the Successive Cancellation List (SCL) algorithm that further improves the performance.

Steganographic coding schemes incorporated with polar codes significantly increase the diversity of steganography. Besides, the computational efficiency of the steganographic schemes is also improved due to the low complexity of the encoding and decoding of polar codes. However, since the polar codes-based steganographic coding always processes bit-by-bit [2,29], the state-of-the-art schemes [7,8,23] still cannot reach the requirements of the practical steganographic application. As the channel polarization process is recursive, the coding procedure under the sub-polarized channels, can be simplified when the type of which is typical [1,17,27]. Therefore, it is possible to improve the overall performance of the coding scheme in steganography.

By establishing sub-polarized steganographic channels, this paper presents a high-performance steganographic codes, named Sub-Polarized Steganographic Codes (SPSC), which is close to the rate-distortion bound. In this scheme, the computation complexity significantly decreases while the embedding efficiency of which is improved. The contributions of this paper are listed as follows.

- Based on the discrete binary symmetric channels, construct polarized and sub-polarized steganographic channels. Define the relationship between the log-likelihood ratio (LLR) and the embedding distortion of cover element.
- Propose an encoding strategy under the sub-polarized steganographic channels and implement a near-optimal steganographic coding scheme along with the existing list decoding algorithm of polar codes.
- Present efficient listed coding schemes for four identified sub-polarized channels, which are denoted as R1, DC, SPC and R0. Construct steganographic codes under these sub-polarized channels.

The rest of this paper is organized as follows. In Sect. 2, we first restate the preliminaries of the optimal embedding theory and channel polarization. The proposed steganographic coding scheme is elaborated in Sect. 3 as well as the construction of sub-polarized steganographic channel. Section 4 gives the experimental results and analysis in detail. Finally, a brief conclusion of this paper is listed in Sect. 5.

## 2 Preliminaries

In this paper, matrices and vectors are written in boldface while sets are shown in swash letters. Without loss of generality, let $\mathbf{x} = (x_1, x_2, \cdots, x_N) \in \{\mathcal{L}\}^n$ and $\mathbf{y} = (y_1, y_2, \cdots, y_N) \in \{\mathcal{I}_i\}^N$ denote the cover and stego sequence, respectively, where $\mathcal{L}$ represents the dynamic range of cover elements and $\mathcal{I}_i \subset \mathcal{L}$ stands for the operation of modification on $x_i$. For instance, $\mathcal{I}_i = \{x_i, \bar{x}_i\}$ is for binary embedding where $\bar{x}_i$ is the least significant bit (LSB) flipped element with respect to $x_i$ and $\mathcal{I}_i = \{x_i - 1, x_i, x_i + 1\}$ for the ternary embedding mode.

Besides, $h(x)$ is the binary quantizer that returns 0 when $x \geq 0$ while returns 1 otherwise. And $\mathcal{P}(x) = x \mod 2$ is defined for extracting LSB of input element. $H_q(x)$ is used for representing $q$-ary entropy function. Typically, the binary entropy function is defined as $H_2(x) = -[x \log_2(x) + (1 - x) \log_2(1 - x)]$.

### 2.1 The Theory of Optimal Steganographic Embedding

For a given cover sequence $\mathbf{x} \in \mathcal{X} \triangleq \{\mathcal{L}\}^n$, the stego sequence can be denoted as $\mathbf{y} \in \{\mathcal{I}_i\}^n \subset \mathcal{X}$ when the modified pattern $\mathcal{I}_i \subset \mathcal{I}$ for $\mathbf{y}$ is defined. In this case, the process of steganography can be defined by modification transition distribution $\pi(\mathbf{y}) \triangleq P(\mathbf{y} \mid \mathbf{x})$.

Under the additive model, if the message $\mathbf{m}$ with length of $|\mathbf{m}|$ to be embedded is given, to obtain the optimal distribution of $\pi(\mathbf{y})$, it is equivalent to solving the PLS problem which is as follows

$$\min_{\boldsymbol{\pi}} E_{\pi}[D] = \sum_{\mathbf{y} \in \mathcal{Y}} \pi(\mathbf{y}) D(\mathbf{y}) \tag{1}$$

$$\text{subject to } H(\boldsymbol{\pi}) = -\sum_{\mathbf{y} \in \mathcal{Y}} \pi(\mathbf{y}) \log_2 \pi(\mathbf{y}) = |\mathbf{m}| \tag{2}$$

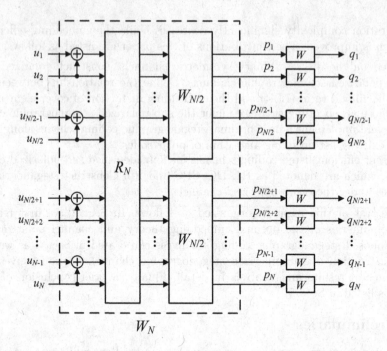

**Fig. 1.** Illustration of recursive construction of polarized channel $W_N$.

where $\boldsymbol{\rho} = (\rho(y_1), \rho(y_2), \cdots, \rho(y_n))$ denotes the pre-defined additive distortion of each element and

$$D(\mathbf{y}) = \sum_{i=1}^{n} \rho(y_i) \tag{3}$$

denotes the distortion under given modification pattern. This optimization problem can be solved with Lagrangian multiplier [12]. And the optimal distribution of modification for each element is

$$\pi_\lambda(y_i) = \frac{\exp[-\lambda\rho(y_i)]}{\sum_{y' \in \mathcal{I}_i} \exp[-\lambda\rho(y')]} \tag{4}$$

where $\lambda > 0$ is a scalar parameter determined by Eq. (2). The distribution above is the best mapping from modification probability to steganographic distortion.

## 2.2 Channel Polarization

Channel polarization [2] is the method for constructing polar codes, the first provable capacity-achieving channel code. For distinguishing them from the notations in steganography, use $\mathcal{P} = \{0, 1\}$ and $\mathcal{Q}$ to represent the input and output alphabets, respectively. When given a B-DMC $W: \mathcal{P} \to \mathcal{Q}$ with transition probabilities $W(q \mid p)$, $p \in \mathcal{P}$, $q \in \mathcal{Q}$, the process of channel polarization is mainly consists of two phases: Channel Combining and Channel Splitting.

After which, $N$ synthesized bit channels $W_N^{(i)}$ are polarized and have symmetric capacity either close to 0 or close to 1 as $N$ approaches infinity. It is proved that the channel capacity can be achieved by transmitting information bits using the noiseless channels [2].

Figure 1 gives the general recursive form of coding procedure of polar codes with the length of $N = 2^n$. Based on which the linear mapping $\mathbf{u} \to \mathbf{p}$, i.e., the encoding procedure of polar codes, is established which can be expressed as $\mathbf{p} = \mathbf{u}\mathbf{G}_N$ where $\mathbf{G}_N$ is the transform matrix that

$$\mathbf{G}_N = \mathbf{B}_N \mathbf{F}_2^{\otimes n}, \quad \mathbf{F}_2 = \begin{bmatrix} 1 & 0 \\ 1 & 1 \end{bmatrix}. \tag{5}$$

$\mathbf{B}_N$ is the bit-reversal permutation matrix consists of the operation $R_N$ and $\otimes$ denotes the Kronecker power.

Denote the set of indices of bit channels with smallest code rate as $\mathcal{A}^c$, the most critical step in polar codes construction is how to determine $\mathcal{A}^c$, i.e., the indices of frozen bits. As for B-DMC, Bhattacharyya parameter $Z$ is generally used as a measure of quality of split bit channels, which can be calculated by

$$Z(W_{2N}^{(2i-1)}) \le 2Z(W_N^{(i)}) - \left[Z(W_N^{(i)})\right]^2, \tag{6}$$

$$Z(W_{2N}^{(2i)}) = \left[Z(W_N^{(i)})\right]^2, \tag{7}$$

$1 \le i \le N$ where the equality holds if and only if $W$ is a binary erasure channel (BEC). And the initial value of which is defined as

$$Z(W_1^{(1)}) = Z(W) = \sum_{q \in \mathcal{Q}} \sqrt{W(q \mid p = 0) \cdot W(q \mid p = 1)}. \tag{8}$$

Thus, the smaller the $Z(W_N^{(i)})$, the more reliable the $W_N^{(i)}$ is. After which a $(N, K, \mathcal{A}^c, \mathbf{u}_{\mathcal{A}^c})$ polar code is specified where $\mathbf{u} = (\mathbf{u}_{\mathcal{A}}, \mathbf{u}_{\mathcal{A}^c})$ denotes the source word and $\mathbf{u}_{\mathcal{A}^c}$ represents the frozen bits of length $N - K$.

## 2.3   Decoding of Polar Codes

Since each coordinate channel $W_N^{(i)}$ are successively used in polarized channels, in the decoding procedure, the unfrozen source word $u_i$ can be calculated by previous estimated source words and received codewords, which is

$$\hat{u}_i = \arg \max_{u_i \in \{0,1\}} W_N^{(i)}(\mathbf{r}, \hat{u}_1, \hat{u}_2, \cdots, \hat{u}_{i-1} \mid u_i). \tag{9}$$

This method is SC decoding algorithm [2]. To simplify the calculation, the LLR of each coordinate channel, which is defined as

$$L_N^{(i)} = \ln \frac{W_N^{(i)}(\mathbf{r}, \hat{u}_1, \hat{u}_2, \cdots, \hat{u}_{i-1} \mid 0)}{W_N^{(i)}(\mathbf{r}, \hat{u}_1, \hat{u}_2, \cdots, \hat{u}_{i-1} \mid 1)} \tag{10}$$

is used for codeword decision [21]. Therefore, the $\hat{u}_i = 0$ if $L_N^{(i)} \leq 0$ otherwise $\hat{u}_i = 1$. The calculation of LLR can be recursively formulated by

$$L_{2N}^{(2i-1)} = F\left(L_N^{(i)}, L_N^{(i+N)}\right), \tag{11}$$

$$L_{2N}^{(2i)} = G\left(L_N^{(i)}, L_N^{(i+N)}, \hat{u}_{2i-1}\right) \tag{12}$$

which further implies that the decision of current codeword strongly depends on the previous estimated bits. $F(\cdot)$ and $G(\cdot)$ are the $f$ and $g$ functions defined in [2] in logarithm domain, which is

$$F(a, b) = 2\tanh^{-1}\left(\tanh\left(\frac{a}{2}\right)\tanh\left(\frac{b}{2}\right)\right), \tag{13}$$

$$G(a, b, \omega) = (1 - 2\omega)a + b. \tag{14}$$

To overcome the errors accumulated in the successive decoding process, in SCL decoder [29], $L$ lists of candidate codewords, i.e., decoding paths, are reserved when estimating each bit as well as path metrics (PM) of each path. For unfrozen bit, each path generates two paths by decoding $\hat{u}_i = 0$ and $\hat{u}_i = 1$ therefore a total $2L$ paths are obtained. The metric of $l$-th path can be updated at the $i$-th decoding bit by

$$\mathrm{PM}_i^l = \sum_{k=1}^{i} \ln\left(1 + \exp\left(-(1 - 2\hat{u}_k)L_N^{(k)}\right)\right) \tag{15}$$

and only $L$ paths with lowest PM are maintained for further decoding. The accumulation of errors is greatly reduced with listed SC decoder and the decoding performance is better improved.

## 3   Steganographic Coding on Sub-Polarized Channel

In this section, the polarized steganographic channel and its sub-channels are established through optimal embedding theory. Besides, the steganographic coding methods under the typical sub-channels are given.

### 3.1   Polarized Steganographic Channel

Under the optimal embedding theory, the steganographic embedding process for each cover element can be simulated as a communication process under a lossy channel [23], which is shown in Fig. 2b where the cover element changes to the stego through the decoding process under the simulated channel. This model, named binary steganographic channel (BSteC), is obviously equivalent to BSC where the modification probability $\pi_\lambda(\bar{x}_i)$ is equal to the crossover probability $p_e$ of BSC. As a result, the steganographic coding is formulated by the problem that, given the received codeword (cover) $\mathbf{x} = (x_1, x_2, \cdots, x_N)$, decoding the

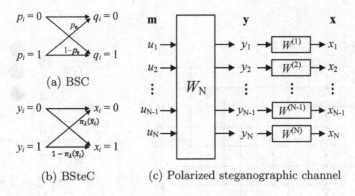

**Fig. 2.** Demonstration of the relationship between communication channel and embedding of steganography, where $W^{(i)}$, $1 \leq i \leq N$ represents BSteC.

codeword before transmitting (stego) **y** of which syndrome (message) is **m** under $N$ independent BSCs with crossover probability $\pi_\lambda(\bar{x}_i)$.

The model of polarized steganographic channel is shown in Fig. 2c where multiple different steganographic channels instead of independent copies are combined and split as the modification pattern of each cover element is independent and different. Under which the steganographer can perform steganographic coding by constructing polar codes with non-uniform channel polarization [24].

However, some prior knowledge, such as embedding distortion, is required to be shared with the extractor for reconstruction of polarized channels, which is nearly impossible. As pointed out in [33], the polar codes constructed for BSC by Eq. (6) and (7) with the equal sign holds still has good performance. Therefore, it can be assumed that all BSteCs are identical and treated as BECs in the construction of polarized steganographic channel. The Bhattacharyya parameter can be used as a metric of channel quality since it is currently the best for BEC and BSC as discussed in [23,33]. Besides, the initial value of Bhattacharyya parameter is discussed in Sect. 4.1.

## 3.2 Successive Cancellation on Polarized Steganographic Channel

According to the recursive form of channel polarization, $N = 2^n$ length polar codes can be represented by a binary tree $T_n(0)$ of depth $n$ [1]. As shown in Fig. 3, each node $T_t(\phi)$ corresponds to a codeword and has a left child $T_{t-1}(2\phi)$ and right child $T_{t-1}(2\phi+1)$. As a result, each sub-tree with a node corresponds to a sub-polar codes which is constructed through a sub-polarized channel.

In the decoding process, SC and SCL decoder sequentially estimates each codeword in a depth-first order. However, partially sub-polar codes have a special form based on the position of frozen bits in the source words, which consists of a special form of codeword. Therefore a limited number of candidate codewords can be directly estimated without recursively calculating all LLRs for decoding all source words. For steganographic codes, the coding scheme under the identified

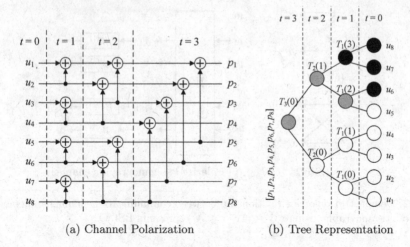

(a) Channel Polarization          (b) Tree Representation

**Fig. 3.** Polar codes and its tree representation with length of 8, where $u_1$ to $u_5$ is assigned as frozen bits. In the binary tree, all white nodes represent frozen bits while all black nodes denote unfrozen bits.

sub-polarized channel can not only reduce the accumulation of distortions in the successive process but also greatly improve computational efficiency.

In this paper, denoted $\mathcal{V}$ as the set of nodes of sub-polar codes, 2-dim tuple $(k_i, S_i) \in \mathcal{V}$ is used to denote a node of polar code with length of $S_i$ of which source word $(p_{k_i}, p_{k_i+1}, \cdots, p_{k_i+S_i-1}) = (u_{k_i}, u_{k_i+1}, \cdots, u_{k_i+S_i-1})\mathbf{G}_{S_i}$. The detailed steganographic coding scheme under the sub-polarized channel is listed in Algorithm 1.

### 3.3  Steganographic Coding Under the Typical Sub-Channel

In this subsection, the corresponding algorithms of steganographic coding will be given under four typical form of sub-polarized channels. The listed successive coding scheme is adopted for better coding efficiency. Based on an important theorem in [18, Théorem 1], the calculation of path metric defined in Eq. (15) can be updated as

$$\text{PM}_i^l = \sum_{k=1}^i \ln\left(1 + \exp\left(-(1 - 2\hat{p}_k) L_1^{(k)}\right)\right) \tag{16}$$

where $(\hat{p}_1, \hat{p}_2, \cdots, \hat{p}_S) = (\hat{u}_1, \hat{u}_2, \cdots, \hat{u}_S)\mathbf{G}_S$. This equation will be adopted further in the steganographic coding procedure. Note that any polarized channel can be represented by the sub-channels below, since the shortest length of codeword is 2 under these sub-channels.

---

**Algorithm 1:** Binary Embedding of Steganography Under the Sub-Polarized Steganographic Channel

---

**Input:** cover $\mathbf{x}$, message $\mathbf{m}$, costs $\rho$ and list size $L$.

**Output:** stego $\mathbf{y}$ and total distortion $D(\mathbf{y})$.

1   calculate Bhattacharyya parameter $Z(W_N^{(i)})$ for each bit channel by Equation (6) and (7) with initial value $Z(W_1^{(1)})$ by Equation (19);

2   get the set of indices $\mathcal{A}^c$ of bit channels with largest metric $Z(W_N^{(i)})$ where $|\mathcal{A}^c| = |\mathbf{m}|$; set the frozen bits of polarized steganographic channels as message $\mathbf{m}$, which is $\mathbf{u}_{\mathcal{A}^c} = \mathbf{m}$;

3   calculate optimal probability of modification of each element $\pi_\lambda(y_i)$ by Equation (4); and further calculate the LLR $L_1^{(i)}$ of each cover element with $\mathcal{P}(\mathbf{x})$ by Equation (10);

4   recursively identify the type of sub-polarized channel $(k_i, S_i)$ and storage 2-dim tuples in the set $\mathcal{V}$;

5   **foreach** *sub node* $(k_i, S_i)$ *in set* $\mathcal{V}$ **do**

6      recursively calculate the LLR $L_N^{(k_i)}$ for $k_i$-th source word;

7      according to the type of sub node, select the corresponding coding scheme in Subsection 3.3 to generate the estimated codeword $\hat{\mathbf{p}}$;

8      obtain the estimated source word $(\hat{u}_{k_i}, \hat{u}_{k_i+1}, \cdots, \hat{u}_{k_i+S_i-1})$ by polar encoding (5);

9      recursively update all partial sums, i.e., intermediate codewords of all coding paths;

10 **end**

11 calculate $\mathcal{P}(\mathbf{y})$ with $\hat{\mathbf{u}}$ by polar encoding (5);

12 get the stego $\mathbf{y} = \mathbf{x} - \mathcal{P}(\mathbf{x}) + \mathcal{P}(\mathbf{y})$; calculate $D(\mathbf{y}) = \sum \rho(y_i)$.

---

**R1 (Rate-1) Channel.** Under the Rate-1 channel denoted as $(k_i, S_i)$, all coordinate channels are used to transmit frozen bits which implies that the message transmission rate is 1. For the steganographic coding, there is only one valid codeword $\mathbf{p}_0 = (u_{k_i}, u_{k_i+1}, \cdots, u_{k_i+S_i-1})\mathbf{G}_{S_i}$ where $u_k \in \mathcal{A}^c$, $k_i \le k < k_i + S$ is set by the message $\mathbf{m}$, thereafter no path splitting occurs. Besides, since the coding procedure is all conducted on frozen bits, the increment of path metrics $\Delta\mathrm{PM}^l$ are directly updated by Eq. (16).

**DC (Dual Candidate) Channel.** The node of polar codes based on DC channel is already discussed in [27, Section IV-B]. Under the DC channels denoted by $(k_i, S_i)$, all source words are determined as frozen bit except the last bit $u_{k_i+S_i-1}$. As a result, there are only two candidate codewords exists, which are $\mathbf{p}_0 = (u_{k_i}, \cdots, u_{k_i+S_i-2}, 0)\mathbf{G}_{S_i}$ and $\mathbf{p}_1 = (u_{k_i}, \cdots, u_{k_i+S_i-2}, 1)\mathbf{G}_{S_i}$. Each path generates two candidate paths whose PM are updated by Eq. (16).

**R0 (Rate-0) Channel.** In Rate-0 node, all source words are unfrozen bits. The maximum likelihood (ML) decision of codeword $p_k, k_i \le k < k_i + S_i$ discussed in [1, Lemma 1] are $\hat{p}_k = h(L_{S_i}^{(k)})$. However, $\hat{p}_k$ is not necessarily the best coding result. The other near-ML decisions have to be obtained for better performance. As for list coder with $L$ paths, the near-ML codes can be obtained by only flipping each codeword in $\hat{p}_k$ the ascending order of corresponding LLR $L_{S_i}^{(k)}$. After each flipping, twice as many near-ML codewords are generated while at most $L$ candidate codewords are reserved based on the path metric increment $\Delta PM^l$. This operation is only performed on the codewords corresponding to the first $L - 1$ smallest LLRs. After that, $L$ paths of candidates will be generated.

**SPC (Single Parity Check) Channel.** For the node constructed by SPC channel denoted by $(k_i, S_i)$, only the first coordinate channel transmits frozen bits, i.e., $u_{k_i} \in \mathcal{A}^c$ and $u_{k_i+1}, u_{k_i+2}, \cdots, u_{k_i+S_i-1} \notin \mathcal{A}^c$. The parity of all source word in this sub-polar codes equals to $u_{k_i}$.

**Theorem 1.** *In SPC channel, the parity of codeword of sub-polar codes satisfies*

$$P = \bigoplus_{k=k_i}^{k_i+S_i-1} p_k = u_{k_i}$$

*where $\oplus$ represents modulo 2 addition.*

*Proof.* We proof this theorem with induction. For any polar codes with $n = 1$, the codeword $(p_1, p_2)$ equals to $(u_1, u_2)\mathbf{G}_2 = (u_1 \oplus u_2, u_2)$. The theorem holds since $p_1 \oplus p_2 = u_1 \oplus u_2 \oplus u_2 = u_1$. Now suppose the theorem stands for polar codes of SPC mode with $n = k$. For $n = k + 1$, denote the source word $\mathbf{u} = (\mathbf{u}_1, \mathbf{u}_2)$ where $\mathbf{u}_1 = (u_1, u_2, \cdots, u_{2^k})$ and $\mathbf{u}_2 = (u_{2^k+1}, \cdots, u_{2^{k+1}})$ are two separated words. Then for $\mathbf{p} = (\mathbf{p}_1, \mathbf{p}_2) = \mathbf{u}\mathbf{B}_{2^{k+1}}\mathbf{F}_2^{\otimes k+1}$ where $\mathbf{p}_1 = (p_1, p_3, \cdots, p_{2^{k+1}-1})$ and $\mathbf{p}_2 = (p_2, p_4, \cdots, p_{2^{k+1}})$, there are

$$\mathbf{p}_1 = (u_1, u_2, \cdots, u_{2^k})\mathbf{B}_{2^k}\mathbf{F}_2^{\otimes k} \oplus (u_{2^k+1}, \cdots, u_{2^{k+1}})\mathbf{B}_{2^k}\mathbf{F}_2^{\otimes k}$$
$$\mathbf{p}_2 = (u_{2^k+1}, \cdots, u_{2^{k+1}})\mathbf{B}_{2^k}\mathbf{F}_2^{\otimes k}$$

where

$$\mathbf{B}_{2^{k+1}} = \mathbf{R}_{2^k}(\mathbf{I}_2 \otimes \mathbf{B}_{2^k}), \quad \mathbf{F}_2^{\otimes k+1} = \begin{bmatrix} \mathbf{F}_2^{\otimes k} & \mathbf{0} \\ \mathbf{F}_2^{\otimes k} & \mathbf{F}_2^{\otimes k} \end{bmatrix}.$$

Since the bit-reversal permutation matrix only changes the order of codewords, the parity of $\mathbf{p}$ can be obviously calculated by codeword $(u_1, u_2, \cdots, u_{2^k})\mathbf{B}_{2^k}\mathbf{F}_2^{\otimes k}$ which is equivalent to the hypothesis above. Therefore, the theorem is valid for all positive integer $n$.

Similar to the node under the Rate-0 channel, the ML decision of codewords of SPC node are $\hat{p}_k = h(L_{S_i}^{(k)})$ if the parity of codewords equals to $u_{k_i}$. Otherwise, flip the codeword with the smallest LLR to satisfy the requirements of channel. The other $L - 1$ near-ML decisions can still be obtained by sequentially flipping the codewords corresponding to the 2nd to $(L - 1)$-th smallest LLRs. Besides, it is also necessary to flip the codeword corresponding to the minimum LLR to ensure the validity of the codewords.

## 4    Experimental Results

In this section, experiments are mainly conducted on binary embedding by various distortion profiles with randomly generated cover elements and messages. The embedding efficiency $e = |\mathbf{m}|/D(\mathbf{y})$ is used for comparing the performance with the state-of-the-art steganographic coding schemes, where $e_\pi = |\mathbf{m}|/E_\pi(D)$ is the theoretical upper bound of embedding efficiency. The throughput of coding, i.e., average number of cover elements processed per seconds, is also investigated to evaluate the computational efficiency. The distortion profile denoted by $\varrho = (\varrho_1, \varrho_2, \cdots, \varrho_N)$ is defined as $\varrho_i = \varrho(i/N)$ [9,10]. The constant, linear and square profile are used for evaluation.

For comparison, STC [10] with sub-matrix height $h = 8, 10, 12, 16$ and SPC [23] with list size $l = 1, 4, 16$ are introduced in simulations. All schemes are implemented in C++ and compiled in MEX executable format[1].

### 4.1    Construction of Polarized Steganographic Channel

As discussed above, the key point of polarized steganographic channel construction is to calculate the initial value of the Bhattacharyya parameter. In this paper, three heuristically defined strategies are discussed, which are

$$\text{Type I}: \quad Z(W_1^{(1)}) = \frac{1}{n}\sum_{i=1}^{n} 2\sqrt{\pi_\lambda(\bar{x}_i)(1 - \pi_\lambda(\bar{x}_i))}, \quad (17)$$

$$\text{Type II}: \quad Z(W_1^{(1)}) = H_2\left(\frac{1}{n}\sum_{i=1}^{n}\pi_\lambda(\bar{x}_i)\right), \quad (18)$$

$$\text{Type III}: \quad Z(W_1^{(1)}) = \frac{1}{n}\sum_{i=1}^{n} H_2(\pi_\lambda(\bar{x}_i)) = \frac{|\mathbf{m}|}{n}. \quad (19)$$

Under the linear and square distortion profile, the embedding efficiency of the three strategies in Eq. (17), (18) and (19) with proposed scheme of list size of 8 are all evaluated for 100 times, which shown in Fig. 4. The strategy of Type I and Type III that outperform STC both achieve near-optimal performance while Type III performs better. In the proposed scheme, we use the strategy in Type III for channel reliability initialization.

### 4.2    Security Evaluation Under Embedding Efficiency Results

The embedding efficiencies of the proposed scheme are simulated for three typical distortion profiles, which is shown in Fig. 5a, 5b, and 5c. For the constant profile, the polar codes-based coding schemes are worse than STC at the small payloads. While for the other profiles, SPC and SPSC both perform closer to the theoretical

---

[1] The complied MEX executable files of SPSC have been uploaded at https://github.com/martin9676/Polarized-Steganographic-Codes/releases

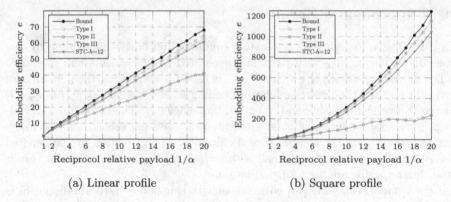

**Fig. 4.** Embedding performance of three different strategy for construction of polarized channel of which list size $L$ is set to 8. Cover elements with length of $2^{20}$ are randomly generated as well as messages.

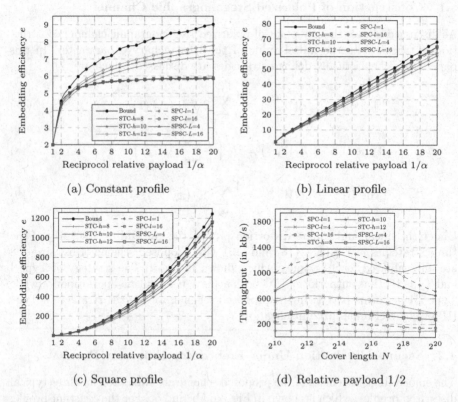

**Fig. 5.** Embedding performance of different steganographic coding schemes. (a)–(c) Embedding efficiency for constant, linear, and square distortion profiles with $N = 2^{20}$ cover elements. (d) Average embedding throughput of different steganographic coding schemes under various cover length. Evaluations above are all conducted with MEX executable version on Intel(R) i7-4790 CPU @ 3.60 GHz.

**Table 1.** The detection error rate (in %) of steganalysis tools where features are extracted by DCTR. STC, SPC and SPSC are compared with J-UNIWARD (JUNI) and UERD on BOSSBase with different quality factors.

| Method | QF = 75 | | | | QF = 90 | | | |
|---|---|---|---|---|---|---|---|---|
| | 0.1 | 0.2 | 0.3 | 0.4 | 0.1 | 0.2 | 0.3 | 0.4 |
| JUNI-STC-$h = 8$ | 0.4099 | 0.2789 | 0.1504 | 0.0618 | 0.4520 | 0.3593 | 0.2434 | 0.1300 |
| JUNI-STC-$h = 12$ | **0.4161** | 0.2909 | 0.1663 | 0.0735 | 0.4564 | 0.3680 | 0.2589 | 0.1470 |
| JUNI-SPC-$l = 4$ | 0.4147 | 0.2935 | 0.1710 | 0.0785 | 0.4564 | 0.3715 | 0.2634 | 0.1533 |
| JUNI-SPC-$l = 16$ | 0.4155 | **0.2953** | 0.1719 | **0.0806** | **0.4566** | **0.3725** | **0.2656** | 0.1558 |
| JUNI-SPSC-$L = 4$ | 0.4153 | 0.2934 | **0.1731** | 0.0786 | 0.4562 | 0.3705 | 0.2650 | 0.1538 |
| JUNI-SPSC-$L = 16$ | 0.4158 | 0.2941 | 0.1726 | 0.0803 | 0.4553 | 0.3719 | 0.2648 | **0.1562** |
| UERD-STC-$h = 8$ | 0.4009 | 0.2701 | 0.1537 | 0.0708 | 0.4425 | 0.3430 | 0.2297 | 0.1258 |
| UERD-STC-$h = 12$ | **0.4069** | 0.2851 | 0.1699 | 0.0831 | 0.4462 | 0.3517 | 0.2434 | 0.1414 |
| UERD-SPC-$l = 4$ | 0.4061 | 0.2865 | 0.1725 | 0.0874 | **0.4471** | 0.3541 | 0.2501 | 0.1487 |
| UERD-SPC-$l = 16$ | 0.4062 | 0.2873 | **0.1754** | **0.0887** | 0.4460 | 0.3555 | **0.2504** | 0.1491 |
| UERD-SPSC-$L = 4$ | 0.4064 | 0.2872 | 0.1732 | 0.0871 | 0.4461 | **0.3570** | 0.2498 | 0.1485 |
| UERD-SPSC-$L = 16$ | 0.4062 | **0.2879** | 0.1744 | 0.0884 | 0.4461 | 0.3561 | 0.2503 | **0.1505** |

bound compared with STC, and SPSC slightly outperforms SPC when the list size is the same. Under the sub-polarized steganographic channel, the proposed coding scheme can reduce the error propagation of recursively coding process and improve the embedding performance.

### 4.3   Security Evaluation Under Image Steganalysis

The anti-steganalysis performance is evaluated by JPEG image steganography with DCTR [19] feature extractor. To conduct the evaluation efficiently, the $F$ function defined in Eq. (11) is approximated when $|a| \leq 10$ or $|b| \leq 10$ by $F(a, b) \approx \text{sgn}(a)\text{sgn}(b) \min \{|a|, |b|\}$, where $\text{sgn}(\cdot)$ is the signum function [26]. Samples sized $512 \times 512$ with quality factor of 75 and 90 are generated from BOSSBase [3] of RAW format. Stego counterparts are embedded by different coding schemes with J-UNIWARD and UERD. As demonstrated in Table 1, compared with classical STC, polar codes-based coding schemes perform slightly better especially when the payload increases. This implies that the proposed scheme has comparable security performance in JPEG image steganography despite that some imprecise approximations are introduced.

### 4.4   Evaluation of the Computational Efficiency

The computational efficiency is first evaluated under the simulated embedding by the throughput with different cover length and fixed payload of 1/4 and 1/2. As demonstrated in Fig. 5d, the throughput of SPSC with list size of 4 is 6–8 times that of STC of $h = 12$ with optimizations of Single Instruction, Multiple

**Table 2.** Evaluation of STC and SPSC on ALASKA2 dataset with 10,000 randomly selected covers. The detection error rate (in %) of steganalysis tools and steganographic coding speed (in second) are evaluated. Evaluation conducted with Intel(R) Xeon(R) CPU E5-2620 v3 @ 2.40 GHz.

|      | Payload $\alpha$ | $h = 8$ | $h = 12$ | $h = 16$ |
|------|------------------|---------|----------|----------|
| STC  | 0.2 bpnzAC | 0.3466/0.07 | 0.3570/1.33 | 0.3625/21.88 |
|      | 0.4 bpnzAC | 0.1636/0.09 | 0.1803/1.39 | 0.1844/22.62 |
|      | Payload $\alpha$ | $L = 1$ | $L = 4$ | $L = 16$ |
| SPSC | 0.2 bpnzAC | 0.3591/0.32 | 0.3608/0.53 | 0.3614/1.40 |
|      | 0.4 bpnzAC | 0.1827/0.33 | 0.1831/0.58 | 0.1851/1.57 |

Data (SIMD) instructions. Meanwhile, with the same list size, the throughput of SPSC is about 1.5 times compared to the SPC. Therefore, the polar codes-based schemes are more capable of performing steganographic coding at higher embedding efficiency with lower computation complexity.

To further evaluate the comprehensive performance of the proposed scheme in practical steganographic applications, stego images are regenerated on ALASKA2 dataset [4] with 10,000 randomly selected samples. Cover images are sized $512 \times 512$ with the quality factor of 85. Steganographic embedding is performed with UERD and statistically analyzed with DCTR feature. To compare the proposed scheme to STC, overall performances including anti-detection performance and embedding speed are both evaluated, and the approximate calculation of the $F$ function is disabled. As shown in Table 2, In the case of similar calculation speed, the steganalysis error rate of SPSC ($L = 16$) is nearly 0.5% higher than that of STC ($h = 12$). When the security performance is equivalent, in the current computing environment, the computing speed of SPSC ($L = 1$) is at least 4 times that of STC ($h = 12$). Therefore, the overall performance of the proposed steganographic coding scheme is far more higher than that of classical STC.

### 4.5   Discussion

The proposed SPSC achieves near-optimal embedding performance that is much better than classical STC by constructing the sub-polarized steganographic channel. It slightly improves security with less consumption of computational resources compared with SPC. As mentioned before, since the Viterbi algorithm is a dynamic programming algorithm that obtains the most likely sequence of hidden states, the computational complexity of STC, which is $\mathcal{O}(2^h N)$, is directly related to the size of the state space $2^h$.

However, the successive cancellation-based algorithm is a greedy algorithm that searches the codeword in the decoding tree with pruning. As a result, its computational complexity is less than $\mathcal{O}(N^2)$ as the calculation of each bit estimation can be completed in logarithmic time. After introducing the list coding strategy, SPC can perform steganographic coding with the computational complexity of $\mathcal{O}(LN \log N)$, where $L$ is the list size.

While the introducing of the sub-polarized steganographic channel further reduces the computational complexity by optimizing the process of bit-by-bit coding. Due to the existence of the partial order relationship of the synthesized channel under the channel polarization [28], multiple bits can be estimated in batches with one or a few estimations, which further improves the coding and computing performance of SPSC, although the theoretical computational complexity is still $\mathcal{O}(LN \log N)$. From the experimental results, SPSC has a slight improvement in security and at least doubles the speed of encoding calculation compared with SPC.

## 5 Conclusion

In this paper, a near-optimal steganographic codes is proposed with constructed polarized and steganographic channel. Four typical sub-polarized channels are introduced which further verifies the effectiveness and the performance of the proposed scheme. Compared with the state-of-the-art steganographic codes, the overall experimental results show that the proposed codes performs embedding with higher efficiency and half of the time consumption, since which it also enables large-scale practical steganography applications. In our future works, the proposed scheme will be further improved with the discovery of sub-polarized channels with more specific patterns.

## References

1. Alamdar-Yazdi, A., Kschischang, F.R.: A simplified successive-cancellation decoder for polar codes. IEEE Commun. Lett. **15**(12), 1378–1380 (2011)
2. Arikan, E.: Channel polarization: a method for constructing capacity-achieving codes for symmetric binary-input memoryless channels. IEEE Trans. Inf. Theory **55**(7), 3051–3073 (2009)
3. Bas, P., Filler, T., Pevný, T.: "Break our steganographic system": the ins and outs of organizing BOSS. In: Filler, T., Pevný, T., Craver, S., Ker, A. (eds.) IH 2011. LNCS, vol. 6958, pp. 59–70. Springer, Heidelberg (2011). https://doi.org/10.1007/978-3-642-24178-9_5
4. Cogranne, R., Giboulot, Q., Bas, P.: Alaska# 2: challenging academic research on steganalysis with realistic images. In: 2020 IEEE International Workshop on Information Forensics and Security (WIFS), pp. 1–5. IEEE (2020)
5. Crandall, R.: Some notes on steganography. Posted on Steganography Mailing List **1998**, 1–6 (1998)
6. Diop, I., Farss, S., Tall, K., Fall, P., Diouf, M., Diop, A.: Adaptive steganography scheme based on LDPC codes. In: 16th International Conference on Advanced Communication Technology, pp. 162–166. IEEE (2014)
7. Diouf, B., et al.: JPEG steganography based on successive cancellation decoding of polar codes. In: 2022 2nd International Conference on Innovative Research in Applied Science, Engineering and Technology (IRASET), pp. 1–6. IEEE (2022)

8. Diouf, B., et al.: Polar coding steganographic embedding using successive cancellation. In: M. F. Kebe, C., Gueye, A., Ndiaye, A. (eds.) InterSol/CNRIA -2017. LNICST, vol. 204, pp. 189–201. Springer, Cham (2018). https://doi.org/10.1007/978-3-319-72965-7_18

9. Filler, T., Judas, J., Fridrich, J.: Minimizing embedding impact in steganography using trellis-coded quantization. In: Media Forensics and Security II, vol. 7541, pp. 38–51. SPIE (2010)

10. Filler, T., Judas, J., Fridrich, J.: Minimizing additive distortion in steganography using syndrome-trellis codes. IEEE Trans. Inf. Forensics Secur. 6(3), 920–935 (2011)

11. Fridrich, J.: Steganography in Digital Media: Principles, Algorithms, and Applications. Cambridge University Press, Cambridge (2009)

12. Fridrich, J., Filler, T.: Practical methods for minimizing embedding impact in steganography. In: Security, Steganography, and Watermarking of Multimedia Contents IX, vol. 6505, pp. 13–27. SPIE (2007)

13. Fridrich, J., Goljan, M., Lisonek, P., Soukal, D.: Writing on wet paper. IEEE Trans. Signal Process. 53(10), 3923–3935 (2005)

14. Fridrich, J., Goljan, M., Soukal, D.: Efficient wet paper codes. In: Barni, M., Herrera-Joancomartí, J., Katzenbeisser, S., Pérez-González, F. (eds.) IH 2005. LNCS, vol. 3727, pp. 204–218. Springer, Heidelberg (2005). https://doi.org/10.1007/11558859_16

15. Guo, L., Ni, J., Shi, Y.Q.: An efficient JPEG steganographic scheme using uniform embedding. In: 2012 IEEE International Workshop on Information Forensics and Security (WIFS), pp. 169–174. IEEE (2012)

16. Guo, L., Ni, J., Su, W., Tang, C., Shi, Y.Q.: Using statistical image model for JPEG steganography: uniform embedding revisited. IEEE Trans. Inf. Forensics Secur. 10(12), 2669–2680 (2015)

17. Hanif, M., Ardakani, M.: Fast successive-cancellation decoding of polar codes: identification and decoding of new nodes. IEEE Commun. Lett. 21(11), 2360–2363 (2017)

18. Hashemi, S.A., Condo, C., Gross, W.J.: A fast polar code list decoder architecture based on sphere decoding. IEEE Trans. Circuits Syst. I Regul. Pap. 63(12), 2368–2380 (2016)

19. Holub, V., Fridrich, J.: Low-complexity features for JPEG steganalysis using undecimated DCT. IEEE Trans. Inf. Forensics Secur. 10(2), 219–228 (2014)

20. Holub, V., Fridrich, J., Denemark, T.: Universal distortion function for steganography in an arbitrary domain. EURASIP J. Inf. Secur. 2014(1), 1–13 (2014). https://doi.org/10.1186/1687-417X-2014-1

21. Leroux, C., Tal, I., Vardy, A., Gross, W.J.: Hardware architectures for successive cancellation decoding of polar codes. In: 2011 IEEE International Conference on Acoustics, Speech and Signal Processing (ICASSP), pp. 1665–1668. IEEE (2011)

22. Li, B., Wang, M., Huang, J., Li, X.: A new cost function for spatial image steganography. In: 2014 IEEE International Conference on Image Processing (ICIP), pp. 4206–4210. IEEE (2014)

23. Li, W., Zhang, W., Li, L., Zhou, H., Yu, N.: Designing near-optimal steganographic codes in practice based on polar codes. IEEE Trans. Commun. 68(7), 3948–3962 (2020)

24. Oliveira, R.M., de Lamare, R.C.: Non-uniform channel polarization and design of rate-compatible polar codes. In: 2019 16th International Symposium on Wireless Communication Systems (ISWCS), pp. 537–541. IEEE (2019)

25. Pevný, T., Filler, T., Bas, P.: Using high-dimensional image models to perform highly undetectable steganography. In: Böhme, R., Fong, P.W.L., Safavi-Naini, R. (eds.) IH 2010. LNCS, vol. 6387, pp. 161–177. Springer, Heidelberg (2010). https://doi.org/10.1007/978-3-642-16435-4_13

26. Ryan, W., Lin, S.: Channel Codes: Classical and Modern. Cambridge University Press, Cambridge (2009)

27. Sarkis, G., Giard, P., Vardy, A., Thibeault, C., Gross, W.J.: Fast polar decoders: algorithm and implementation. IEEE J. Sel. Areas Commun. **32**(5), 946–957 (2014)

28. Schürch, C.: A partial order for the synthesized channels of a polar code. In: 2016 IEEE International Symposium on Information Theory (ISIT), pp. 220–224. IEEE (2016)

29. Tal, I., Vardy, A.: List decoding of polar codes. IEEE Trans. Inf. Theory **61**(5), 2213–2226 (2015)

30. Westfeld, A.: High capacity despite better steganalysis (F5—a steganographic algorithm). In: Moskowitz, I.S. (ed.) IH 2001. LNCS, vol. 2137, pp. 289–302. Springer, Heidelberg (2001). https://doi.org/10.1007/3-540-45496-9_21

31. Zhang, R., Sachnev, V., Botnan, M.B., Kim, H.J., Heo, J.: An efficient embedder for bch coding for steganography. IEEE Trans. Inf. Theory **58**(12), 7272–7279 (2012)

32. Zhang, R., Sachnev, V., Kim, H.J.: Fast BCH syndrome coding for steganography. In: Katzenbeisser, S., Sadeghi, A.-R. (eds.) IH 2009. LNCS, vol. 5806, pp. 48–58. Springer, Heidelberg (2009). https://doi.org/10.1007/978-3-642-04431-1_4

33. Zhao, S., Shi, P., Wang, B.: Designs of Bhattacharyya parameter in the construction of polar codes. In: 2011 7th International Conference on Wireless Communications, Networking and Mobile Computing, pp. 1–4. IEEE (2011)

# High-Capacity Adaptive Steganography Based on Transform Coefficient for HEVC

Lin Yang[1], Rangding Wang[1], Dawen Xu[2]([envelope]), Li Dong[1], Songhan He[1], and Fan Liu[2]

[1] Faculty of Electrical Engineering and Computer Science, Ningbo University, Ningbo, China
{2011082340,wangrangding,dongli,2111082349}@nbu.edu.cn
[2] School of Cyber Science and Engineering, Ningbo University of Technology, Ningbo, China
dawenxu@126.com

**Abstract.** HEVC video is one of the most popular carriers for steganography. The existing transform coefficient-based HEVC steganography algorithms usually modify the coefficients of the candidate blocks to prevent distortion drift. Nevertheless, the embedding capacity is relatively small due to the strict candidate block selection rule, and embedding distortion is accumulated within group of pictures (GOP). In this paper, a novel transform coefficient-based steganography for HEVC is proposed to enlarge embedding capacity and reduce visual degradation. First, the visual distortion and GOP distortion are analyzed to elaborate the embedding influence of different cover coefficients. Next, different cover coefficients are assigned different costs. Besides, the modification in non-zero coefficients of $4 \times 4$ TUs in P-frames is explored to enhance embedding capacity. Moreover, by introducing a new evaluation indicator, it is verified the proposed algorithm can preserve less visual degradation while embedding more secret messages. Experimental results show that the proposed algorithm outperforms the competing methods in terms of visual quality, embedding capacity and anti-steganalysis performance.

**Keywords:** HEVC · High-capacity steganography · Transform coefficient

## 1 Introduction

Steganography is one of the most important branches of data hiding, and it aims to embed secret messages into carriers in an imperceptible way. With the rapid development of digital devices, the transmission and service based on video media have become extensive. HEVC is one of the mainstream video compression standards. Due to more coding information can be modified, it is very suitable for video steganography. As a result, HEVC steganography is one of the hot directions of information hiding.

© The Author(s), under exclusive license to Springer Nature Switzerland AG 2023
X. Zhao et al. (Eds.): IWDW 2022, LNCS 13825, pp. 20–34, 2023.
https://doi.org/10.1007/978-3-031-25115-3_2

Generally speaking, according to the different embedding domain, video steganography can be divided into two types: the spatial domain video steganography and the compressed domain video steganography. The first type modifies the pixel values in the spatial domain, and the second type refers to embedding secret messages by modifying the coding information during the video compression process. Since videos are mainly stored and transmitted in compressed format, the second type of video steganography has more practical application value. The coding information used for steganography includes: intra prediction modes (IPMs) [1–4], motion vectors (MVs) [5–8], transform coefficients [9–15] and partition modes of prediction unit (PU) [16,17]. There are several advantages for selecting transform coefficients as embedding covers. For example, the signal energy of embedded messages will be dispersed to all elements in the spatial domain during the discrete cosine transform (DCT) and inverse DCT process, which enhances the security of steganography behavior. Moreover, transform coefficients are a type of coding information with a large weight in video stream, which ensures a large embedding capacity. Therefore, modifying transform coefficients is a popular way in video steganography.

As far as selecting transform coefficients as carriers is concerned, there are many works have been operated in H.264/AVC. In order to cope with the distortion drift problem, Ma et al. [9] modified the paired-coefficients of some specific blocks to embed secret messages. On the basis of [9], Xue et al. [10] used syndrome trellis codes (STCs) to reduce the intra-block distortion caused by modulating transform coefficients in I-frames. Chen et al. [11] proposed an adaptive steganography by taking three factors, i.e., texture features, motion characteristics and position of the frame, into consideration and used STCs to minimize the total distortion. Chen et al. [12] analyzed the intra-block distortion, inter-block distortion and inter-frame distortion, then proposed an adaptive steganography method named DDCA (distortion-drift cost assignment) and achieved better performance in terms of visual quality.

However, due to the difference in the transformation process between AVC and HEVC, transform coefficient-based steganography in AVC cannot be directly transferred into HEVC. So, the research on transform coefficient-based HEVC steganography is still in its infancy. Liu et al. [13] proposed a transform coefficient-based steganography scheme for HEVC. In order to prevent the distortion drift, they selected the $4 \times 4$ blocks without vertical distortion propagation or horizontal distortion propagation as candidate blocks, and the triple coefficient group was used for distortion compensation. Chang et al. [14] proposed an error propagation-free algorithm for HEVC. After analyzing the relationship of IPMs in surrounding blocks, they located the transform coefficients which can be modified without propagating errors to other blocks. On the basis of [13] and [14], Zhou et al. [15] proposed a BLB (block-based) distortion model to minimize the embedding distortion. They applied method [14] in small blocks with size of $4 \times 4$ and $8 \times 8$ and used STC to minimize the distortion in large blocks with size of $16 \times 16$ and $32 \times 32$.

Although these methods can completely eliminate the distortion drift in I-frames, they did not fully exploit the advantages of HEVC: (1) The selection rule for candidate blocks is very strict, resulting in a small embedding capacity, which loses the superiority of large embedding capacity in HEVC. (2) In order to eliminate inter-block distortion, the intra-block distortion is increased by the distortion compensation algorithm. (3) The influence of the subsequent P-frames is ignored, which causes the large accumulative distortion in the GOP.

In order to overcome the above problems and utilize the characteristic of large embedding capacity in HEVC, we propose a high-capacity adaptive steganography. The contributions of this work can be summarized as follows:

- A novel distortion function is proposed according to the analysis of visual distortion and GOP distortion.
- Explore the modification in non-zero coefficients of $4 \times 4$ transform units (TUs) in P-frames and achieve high-capacity steganography while preserving less visual degradation.
- A new evaluation indicator is used to measure visual degradation for different capacity.

The rest of this paper is organized as follows. The embedding distortion is analyzed in Sect. 2. In Sect. 3, the proposed algorithm is described. Section 4 represents the extensive experiments and analysis. At last, Sect. 5 concludes this paper.

## 2   Analysis of Embedding Distortion

In this section, the embedding distortion is analyzed from two perspectives of visual distortion and GOP distortion, then the influence of modifying coefficients in intra-frame distortion and inter-frame distortion is concluded.

### 2.1   Embedding Distortion of Visual Degradation

Video visual quality is often measured by the Peak Signal to Noise Ratio (PSNR) indicator, and the visual degradation of stego video is defined by the difference between the PSNR of un-stego video and stego video. For better understanding, we assume the pixel depth of each frame is 8-bit, then the PSNR can be expressed as:

$$PSNR = 10 \cdot \lg \frac{255^2 \cdot MN}{\sum_{x=0}^{M-1} \sum_{y=0}^{N-1} |f^o(x,y) - f^c(x,y)|^2} \tag{1}$$

$$f^c(x,y) = f^c_{pre}(x,y) + f^c_{res}(x,y), \tag{2}$$

where $M$ and $N$ represent the width and height of the video and $f^o(x,y)$ and $f^c(x,y)$ represent the pixels in $(x,y)$ position of the original video and the compressed video. $f^c(x,y)$ can be obtained by the sum of prediction value $f^c_{pre}(x,y)$ and residual value $f^c_{res}(x,y)$ as Eq. (2).

The visual quality degradation can be expressed by the difference between the PSNR of un-stego videos and stego videos:

$$\Delta PSNR = PSNR^u - PSNR^s$$

$$= 10 \cdot \lg \frac{255^2 \cdot MN}{\sum_{x=0}^{M-1} \sum_{y=0}^{N-1} |f^o(x,y) - f^u(x,y)|^2}$$

$$- 10 \cdot \lg \frac{255^2 \cdot MN}{\sum_{x=0}^{M-1} \sum_{y=0}^{N-1} |f^o(x,y) - f^s(x,y)|^2} \tag{3}$$

$$= 10 \cdot \lg \frac{\sum_{x=0}^{M-1} \sum_{y=0}^{N-1} |f^o(x,y) - f^s(x,y)|^2}{\sum_{x=0}^{M-1} \sum_{y=0}^{N-1} |f^o(x,y) - f^u(x,y)|^2}$$

$$= 10 \cdot \lg \frac{\sum_{x=0}^{M-1} \sum_{y=0}^{N-1} |f^o(x,y) - f^s_{pre}(x,y) - f^s_{res}(x,y)|^2}{\sum_{x=0}^{M-1} \sum_{y=0}^{N-1} |f^o(x,y) - f^u(x,y)|^2}.$$

Considering that the coefficient-based steganographic algorithms usually modify coefficients after quantization, the prediction values in stego coding process and un-stego coding process are the same, i.e., $f^s_{pre}(x,y) = f^u_{pre}(x,y)$. As a result, Eq. (3) can be simplified as:

$$\Delta PSNR = 10 \cdot \lg \frac{\sum_{x=0}^{M-1} \sum_{y=0}^{N-1} |f^o(x,y) - f^u_{pre}(x,y) - f^s_{res}(x,y)|^2}{\sum_{x=0}^{M-1} \sum_{y=0}^{N-1} |f^o(x,y) - f^u(x,y)|^2}$$

$$= 10 \cdot \lg \frac{\sum_{x=0}^{M-1} \sum_{y=0}^{N-1} |f^u_{res}(x,y) - f^s_{res}(x,y)|^2}{\sum_{x=0}^{M-1} \sum_{y=0}^{N-1} |f^o(x,y) - f^u(x,y)|^2}. \tag{4}$$

It can be seen from Eq. (4) that the value of $\Delta PSNR$ is proportional to the square sum of the difference between the un-stego residual and stego residual. As for the mainstream distortion compensation algorithms [13–15], it is required to modify three coefficients at most to embed 1-bit binary secret message to prevent the distortion drift, which leads to the large change of stego residual, increases the $\Delta PSNR$ and reduces the visual quality.

Take the distortion compensation method in I-frames and Least Significant Bit (LSB) method in P-frames as an example, and each method is used to modify the coefficient $q_{(0,0)}$ in $(0,0)$ position with $+1$ disturbance. In order to eliminate the inter-block distortion, the distortion compensation method needs to compensate the corresponding coefficient group, so the coefficients $q_{(0,2)}$ and $q_{(0,3)}$ in $(2,0)$ and $(3,0)$ positions are modulated as well. The modified pixels are represented as residual values, which are the difference between pixel values and predicted values as Eq. (2).

It can be seen from Fig. 1 that the residual variation of distortion compensation method is concentrated in the third column, while the residual variation of LSB method is distributed throughout the block. It can be calculated from subgraphs (b)(d) and subgraphs (f)(h) that the sum of squares of the differences between the original residual and stego residual are 309 and 104 respectively.

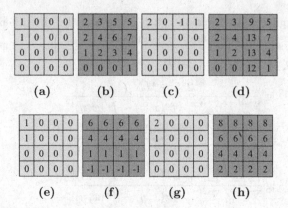

Fig. 1. The influence of pixels in the spatial domain after using distortion compensation method and LSB method. Subgraphs in the first row are the original coefficient matrix, original residual matrix, stego coefficient matrix and stego residual matrix of distortion compensation method in I-frame. Subgraphs in the second row are the matrices of LSB method in P-frame.

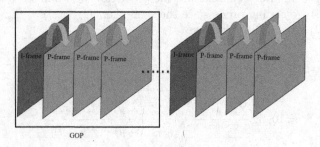

Fig. 2. The illustration for GOP.

Since the distortion compensation algorithm eliminates the inter-block distortion in I-frames and there is no inter-block distortion in P-frames, the former algorithm will reduce the visual quality more than the parity-mapping modification in P-frames when embedding 1-bit binary message.

## 2.2   Embedding Distortion of GOP

In this part, the inter-frame distortion propagation will be analyzed from the perspective of GOP. In HEVC coding standard, video sequences are usually divided into several small GOP to enhance the compression efficiency, as shown in Fig. 2. Every GOP is composed of one I-frame as the starting frame and several P-frames. The first I-frame is encoded independently while the rest P-frames are predicted by the former frame, moreover, the GOP coding is independent and does not affect each other.

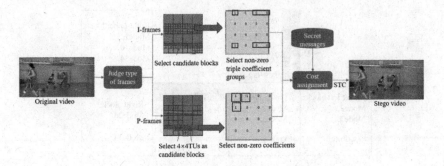

**Fig. 3.** Framework of the proposed steganography algorithm.

Assume there are $n$ frames in a GOP and the first frame is I-frame and the rest frames are P-frames. So, the embedding influence of each frame can be indicated as:

$$\begin{cases} E(0) = D_e(0) \cdot (1 + IDPR)^{n-1} \\ E(1) = D_e(1) \cdot (1 + IDPR)^{n-2} \\ \qquad \vdots \\ E(n-1) = D_e(n-1) \end{cases} \tag{5}$$

where $E(i)$ represents the embedding influence of $i$-$th$ frame, $D_e(i)$ represents the embedding distortion of $i$-$th$ frame and $IDPR$ represents the inter-frame distortion propagation rate.

Since $IDPR$ is in the range of $[0, 1]$, it can be seen from Eq. (5) that the embedding influence of each frame not only includes the embedding distortion but also includes the influence on the inter-frame distortion of the subsequent frames in the GOP. The embedding influence decreases with the increase of the position of video frame in the GOP, and the influence of the inter-frame distortion is smaller for the later video frames. For the distortion compensation algorithm, since it modifies the coefficients in I-frames, the distortion will be accumulated in the whole GOP during the inter-frame prediction process, thus reducing the video visual quality.

## 3    Proposed HEVC Steganography

In this section, according to the analysis in Sect. 2, a high-capacity adaptive steganography for HEVC is proposed. By modeling the GOP distortion, all non-zero coefficients in the blocks without vertical distortion propagation and horizontal distortion propagation in I-frames and non-zero coefficients in $4 \times 4$ TUs in P-frames are chosen as covers. The cost of each cover is calculated through the proposed cost function and STC is utilized for the minimization distortion embedding. The framework of the proposed algorithm is shown in Fig. 3.

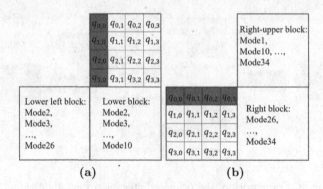

**Fig. 4.** Selection rule of candidate blocks in I-frames. Subgraph (a) represents the selection rule for vertical error propagation-free blocks, whose IPMs of left-lower block and lower block are in the range of [2, 26] and [2, 10]. Subgraph (b) represents the selection rule of horizontal error propagation-free blocks.

## 3.1 Selection of Cover Coefficients

In order to make better use of the characteristics of large embedding capacity in HEVC, we explore the modification of non-zero coefficients in P-frames. Considering the human visual system (HVS) is sensitive to the slight distortion in large flat areas, the $4 \times 4$ TUs within $8 \times 8$ CUs are selected as candidate blocks in P-frames. As for I-frames, the blocks without vertical distortion propagation and horizontal distortion propagation are selected as candidate blocks, which is shown in Fig. 4 and detailed principles and proofs of these two kinds of candidate blocks can refer to work [13]. Since the modification in all zero coefficients will bring large visual distortion, thus the non-zero coefficients in each candidate blocks are selected as cover coefficients in this paper.

As for the candidate blocks in P-frames, all the non-zero coefficients from $q_{(0,0)}$ to $q_{(3,3)}$ are extracted to form the cover sequence. However, for the candidate blocks in I-frames, only the non-zero coefficients in the first column of vertical error propagation-free blocks and the first row of horizontal error propagation-free blocks are extracted as the cover sequence, which are colored as red in the background.

## 3.2 Proposed Cost Function

According to the analysis in Sect. 2, it is known that the visual degradation caused by modifying transform coefficients is proportional to the square sum of the difference between the un-stego residual and stego residual, and it is transferred to the subsequent frames in GOP resulting in cumulative distortion. As a result, the cost of each cover coefficient not only includes the intra-frame distortion but also the influence of inter-frame distortion propagation. Therefore, the cost of each cover $\rho(x_i, y_i)$ is defined as:

$$\rho(x_i, y_i) = \left( \sum_{m=0}^{3} \sum_{n=0}^{3} |res^u(m,n) - res^s(m,n)|^2 \right) \cdot (1 + IDPR)^{f_G - 1 - f\%f_G}, \quad (6)$$

where $res^u(m,n)$ and $res^s(m,n)$ represents the un-stego residual and simulated stego residual in $(m,n)$ position of the corresponding $4 \times 4$ TU block, $f_G$ is the number of total frames in the GOP. $f$ is the index of current frame, which starts from 0. The symbol $\%$ indicates modulo operation. Take the coding structure of *IPPP* as an example, when embedding in the third P-frame, that is $f = 3$, thus $f_G - 1 - f\%f_G = 0$, which means the embedding influence in this frame only includes intra-frame distortion and no influence of inter-frame distortion.

In addition, assume the cover sequences obtained from each frame are: $\boldsymbol{X}^I = \{x_0^I, x_1^I, ..., x_{n_0}^I\}$, $\boldsymbol{X}^{P_1} = \{x_0^{P_1}, x_1^{P_1}, ..., x_{n_1}^{P_1}\}$, $\boldsymbol{X}^{P_2} = \{x_0^{P_2}, x_1^{P_2}, ..., x_{n_2}^{P_2}\}$ and $\boldsymbol{X}^{P_3} = \{x_0^{P_3}, x_1^{P_3}, ..., x_{n_3}^{P_3}\}$ where $n_0$, $n_1$, $n_2$ and $n_3$ are the number of total covers extracted from each frame. Then, the original cover sequence $\boldsymbol{X}$ and stego sequence $\boldsymbol{Y}$ can be expressed as:

$$\begin{aligned} \boldsymbol{X} &= \boldsymbol{X}^I + \boldsymbol{X}^{P_1} + \boldsymbol{X}^{P_2} + \boldsymbol{X}^{P_3} \\ &= \{x_0^I, ..., x_{n_0}^I, x_0^{P_1}, ..., x_{n_1}^{P_1}, x_0^{P_2}, ..., x_{n_2}^{P_2}, x_0^{P_3}, ..., x_{n_3}^{P_3}\} \end{aligned} \quad (7)$$

$$\boldsymbol{Y} = \{y_0^I, ..., y_{n_0}^I, y_0^{P_1}, ..., y_{n_1}^{P_1}, y_0^{P_2}, ..., y_{n_2}^{P_2}, y_0^{P_3}, ..., y_{n_3}^{P_3}\}. \quad (8)$$

Let $D(\boldsymbol{X}, \boldsymbol{Y})$ represents the total distortion of sequence $\boldsymbol{X}$ changed into sequence $\boldsymbol{Y}$ after embedding secret messages $m$, then $D(\boldsymbol{X}, \boldsymbol{Y})$ can be expressed as:

$$D(\boldsymbol{X}, \boldsymbol{Y}) = \sum_{i=0}^{n_0 + n_1 + n_2 + n_3} \rho(x_i, y_i). \quad (9)$$

Then, STC is used to embed secret messages $m$ and calculate the minimization embedding distortion. Detailed introduction and implementation of STC can be found in [18].

Finally, the secret messages $m$ will be embedded into the cover sequence $\boldsymbol{X}$ under limited payload channel, while maintaining small GOP distortion and preserving high visual quality.

### 3.3   The Practical Implementation

The process of the proposed algorithm is shown in Fig. 3 and the practical implementation of the proposed algorithm is introduced as follows:

**Embedding Process.** Firstly, the original YUV sequence is encoded to obtain the optimal coding information of each block, such as IPMs and transform coefficients. Secondly, all non-zero coefficients are extracted based on the candidate block selection in Sect. 3.1, and all extracted coefficients in the same GOP are spliced into a cover sequence. Thirdly, cost of each cover is calculated by Eq. (6) and STC is used for embedding binary secret messages. Next, distortion compensation algorithm is used for the coefficients extracted from I-frames. At last, the original YUV sequence is encoded again using the stego coefficients to obtain the stego video.

**Extraction Process.** As for extraction, the receiver needs to decode the stego video to get the coding information of each block and extract the non-zero coefficients based on the candidate block selection rule in I-frames and P-frames. Finally, all the extracted coefficients in the same GOP are combined into stego sequence and decoded with STC to get secret messages.

## 4   Experimental Results

In this section, the coding configuration is introduced, and the comparison in visual quality, embedding capacity, bit rate increase and anti-steganalysis performance are represented to demonstrate the superiority of the proposed algorithm.

### 4.1   Experimental Setup

**The Configuration of Video Coding.** The proposed steganographic algorithm is implemented in HEVC reference software HM 16.15, and the original configuration *encoder_lowdelay_P_main.cfg* is used in this paper. The coding structure is *IPPP* and the quantization parameter (QP) is set to 24, and the rest of configuration is as same as the original one. The limited payload in this paper is 0.5 bpc (bit per cover) and *IDPR* is 0.5. All the experiments and comparisons are tested under the above configuration.

In order to demonstrate the superior compression capability of HEVC for high-definition (HD) videos, 5 HD HEVC standard test sequences (*Basketball-Drive.yuv, BQTerrace.yuv, Cactus.yuv, ParkScene.yuv, Kimono.yuv*) are used in this paper, each sequence is encoded with 200 frames. Besides, all video sequences are un-compressed and stored in 4:2:0 YUV color space.

**The Comparison Algorithms.** In this paper, the proposed algorithm is compared with the transform coefficient-based HEVC steganography [13,14] and PU-based HEVC steganography [17]. The candidate block selection rule in [13] is as same as the proposed algorithm in I-frames, and the non-zero coefficients are selected as cover coefficients as well. As for method [14], the non-zero coefficients of $8 \times 8$ TUs, which have no inter-block distortion drift, are also modified. So, the embedding capacity is improved. For comparison, all the cover coefficients are embedded in method [13] and [14]. Due to a large number of PU partition modes in HEVC videos, the latest PU-based steganography [17] is used for comparison as well. For method [17], all the cover PU partition modes in the first layer are embedded and the limited payload in the second layer is set to 0.5 bpc as well.

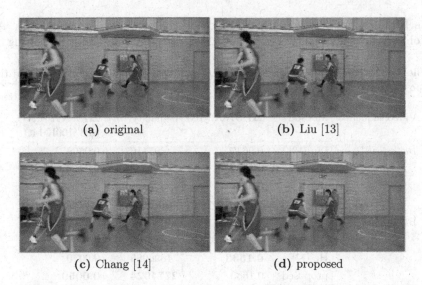

(a) original                        (b) Liu [13]

(c) Chang [14]                      (d) proposed

**Fig. 5.** The comparison results of subjective visual quality with different steganography.

**The Evaluation Indicators.** In this paper, we will compare the proposed algorithm with other HEVC steganography in terms of PSNR, embedding capacity, bit rate increase and steganalysis accuracy. In general, the comparison between different steganography is carried out with the same embedding capacity. However, this cannot demonstrate the capacity advantage of the proposed algorithm. As a result, two indicators, $\Delta PSNR\text{-}NC$($\Delta PSNR$ normalized with 100K bit) and $CPF$, are used to demonstrate the performance of proposed algorithm in terms of visual quality degradation and embedding capacity. These two indicators are defined as:

$$\Delta PSNR\text{-}NC = \frac{PSNR_{un\text{-}stego} - PSNR_{stego}}{capacity} \times 10^5 (\text{dB/100K bit}) \qquad (10)$$

$$CPF = \frac{capacity}{frames}(\text{bit/frame}), \qquad (11)$$

where $PSNR_{un\text{-}stego}$ and $PSNR_{stego}$ represent the PSNR of the un-stego video and stego video, $capacity$ represents the embedding capacity, $frames$ represents the total frame number of the video. It can be observed that the physical meaning of $\Delta PSNR\text{-}NC$ and $CPF$ represent the PSNR loss per embedded 100K bit and embedding capacity in per frame respectively, which are reasonable and realistic.

## 4.2   Performance of Visual Quality

In this part, the performance of visual quality is measured by subjective visual quality and objective visual quality. For subjective visual quality, the first I-frames of the sequence *BasketballDrive* are displayed to represent the imperceptibility from HVS. Since method [17] embeds messages into P-frames rather than I-frames, only method [13] and method [14] are used for comparison. As for objective visual quality, PSNR and $\Delta PSNR\text{-}NC$ are used for measuring.

**Table 1.** The comparison of objective visual quality with method [13,14,17] and the proposed method.

| Video sequence | Method | $\Delta PSNR$ (dB) | Capacity(bit) | $\Delta PSNR - NC$ (dB/100K bit) |
|---|---|---|---|---|
| *BasketballDrive* | Liu [13] | 0.0647 | 97116 | 0.0666 |
| | Chang [14] | 0.1382 | 157023 | 0.0880 |
| | He [17] | 0.0977 | 433853 | 0.0225 |
| | proposed | **0.0610** | **656626** | **0.0093** |
| *BQTerrace* | Liu [13] | 0.1544 | 260777 | 0.0592 |
| | Chang [14] | 0.4529 | 448790 | 0.1009 |
| | He [17] | **0.1333** | 1896985 | 0.0070 |
| | proposed | 0.1633 | **2774924** | **0.0059** |
| *Cactus* | Liu [13] | **0.0839** | 173410 | 0.0484 |
| | Chang [14] | 0.1383 | 216057 | 0.0640 |
| | He [17] | 0.1109 | 1255098 | 0.0088 |
| | proposed | 0.0885 | **1686885** | **0.0052** |
| *Kimono* | Liu [13] | 0.1246 | 198840 | 0.0627 |
| | Chang [14] | 0.1931 | 239527 | 0.0806 |
| | He [17] | **0.0917** | 899871 | 0.0083 |
| | proposed | 0.1029 | **1293749** | **0.0080** |
| *ParkScene* | Liu [13] | **0.0394** | 17787 | 0.2215 |
| | Chang [14] | 0.0455 | 27858 | 0.1633 |
| | He [17] | 0.0868 | **857575** | **0.0101** |
| | proposed | 0.1086 | 104784 | 0.1036 |

The comparison results of subjective visual quality are shown in Fig. 5. It can be seen that the distortion is hardly detectable from HVS. As a result, the proposed algorithm does not cause a large visual quality degradation in subjective evaluation.

As for objective visual quality, the results are shown in Table 1, and the best results are marked in bold. It can be seen from Table 1 that compared with method [13,14] and [17], the proposed algorithm embeds more messages while preserving less PSNR loss in most test sequences. Take the *BasketballDrive* sequence as example, although the proposed method embeds 559510, 499603 and 222773 bits more messages compared with method [13,14] and [17], the PSNR loss is 0.0037, 0.0772 and 0.0367 dB less than them. Considering the different embedding capacity of each algorithm, we calculate $\Delta PSNR\text{-}NC$ for them. In the first four sequences, the $\Delta PSNR\text{-}NC$ of the proposed method is less than method [13,14] and [17], which means the proposed method can preserve less average visual quality degradation. Nevertheless, the $\Delta PSNR\text{-}NC$ of the proposed method is larger than the $\Delta PSNR\text{-}NC$ of method [17] in the last sequence. Theoretically, this is because the *ParkScene* sequence has a flatter frame with less variation in video motion, resulting in fewer $4 \times 4$ TUs in P-frames and reducing the embedding capacity. Moreover, as mentioned in [17], the stego videos generated by PU-based steganography will not be affected significantly. Therefore, the $\Delta PSNR\text{-}NC$ of method [17] is less than the $\Delta PSNR\text{-}NC$ of the proposed method in the *ParkScene* sequence.

### 4.3   Performance of Embedding Capacity

In this part, the embedding capacities of each algorithm are compared. Considering method [13] and [14] extract cover coefficients from I-frames, method [17] extracts cover PU partition modes from P-frames, while the proposed method extracts cover coefficients from I-frames and P-frames, so in order to compare fairly, the indicator $CPF$ is used to measure the average capacity, and the capacity of the proposed method refers to the embedding capacity under the limited payload. The compared results are shown in Fig. 6.

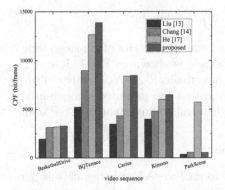 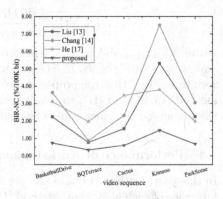

**Fig. 6.** The comparison results of different steganography in $CPF$.

**Fig. 7.** The comparison results of different steganography in $BIR\text{-}NC$.

As shown in Fig. 6, $CPF$ of method [13] is the lowest, which indicates method [13] has the smallest embedding capacity among these methods. This is because method [13] only selects the coefficients of some $4 \times 4$ blocks in I-frames as covers, and in order to cope with distortion drift, the selection rule of candidate blocks is very strict, which sacrifices a large amount of embedding capacity. Compared with method [13], method [14] adds the error propagation-free algorithm for $8 \times 8$ blocks, so the average $CPF$ increases by 1365.30 (bit/frame). As for method [17], since all the PU partition modes except $2N \times 2N$ are selected as covers, the $CPF$ of method [17] is larger than method [13] and [14]. When it comes to the proposed method, the improvement of $CPF$ is significant compared with method [13] and [14], and in the most sequences, the $CPF$ of the proposed

**Table 2.** The classification accuracy results (%) of different steganography.

| Steganalysis | Liu [13] | Chang [14] | Proposed |
|---|---|---|---|
| Zhai [19] | 50.02 | 50.06 | **49.07** |
| Wang [20] | 62.89 | 69.18 | **55.37** |

method is larger than method [17]. However, in the *ParkScene* sequence, the *CPF* of method [17] exceeds that of the proposed method. In theory, due to the flat frame and low texture complexity in the *ParkScene* sequence, coding blocks with large size, e.g., $16 \times 16$ and $32 \times 32$, are used for compression in I-frames and P-frames. Therefore, the embedding capacity of method [13,14] and the proposed is relatively small, i.e., the *CPF* is relatively small.

In addition, the steganographic influence for coding efficiency is tested as well, and the indicator *BIR-NC* (BIR normalized with 100K bit) is used for measuring the influence of coding efficiency. The indicators BIR and *BIR-NC* are defined as:

$$BIR = \frac{bit_{stego} - bit_{un\text{-}stego}}{bit_{un\text{-}stego}} \times 100\% \tag{12}$$

$$BIR\text{-}NC = \frac{BIR}{capacity} \times 10^5 (\%/100K \text{ bit}), \tag{13}$$

where $bit_{un\text{-}stego}$ and $bit_{stego}$ represent the total coding bits of un-stego video and stego video, the *BIR-NC* of each method is shown in Fig. 7. It can be seen from Fig. 7 that the proposed method has smaller *BIR-NC* compared with method [13,14] and [17], so it can be inferred that the proposed algorithm has less impact on the video coding efficiency when embedding capacity is the same.

### 4.4   Performance of Anti-steganalysis

In this part, method [19] and [20] are used to analyze the anti-steganalysis performance of each algorithm. Method [19] proposed a universal feature to capture the steganography traces from multiple feature domain, and method [20] used the spatial domain variation caused by the distortion compensation method for analysis, which is transplanted from H.264/AVC detection to HEVC detection in this paper. In order to compare the performance of anti-steganalysis fairly, only the transform coefficient-based HEVC steganography [13,14] are used for comparison.

All sequences are used to generate the dataset, each sequence is divided into small sub-sequences with 8 frames, and 125 sub-sequences are obtained. LibSVM toolbox is used to classify the extracted features. The proportion of training set and testing set is 1:1, all the data are the average results of 100 repetitions after randomly segmenting dataset. The final results are shown in Table 2.

As for method [19], the classification accuracy of each method is nearly 50%, which means all three steganographic methods can resist this kind of detection. It is because method [19] uses the variation of PU partition mode and

MVs to extract features, however, all three methods do not modify them. As for method [20], method [13] and [14] modify the transform coefficients in I-frames heavily, which leaves obvious traces for detection. Nevertheless, the proposed method regards the coefficients in the same GOP as a cover sequence, and models the GOP distortion for minimization embedding with STC. So, compared with method [13] and [14], the modification in I-frames is much fewer, which diminishes the steganographic traces in I-frames.

# 5   Conclusion

In this paper, a novel high-capacity adaptive steganography scheme for HEVC is proposed. First, the visual distortion and GOP distortion are analyzed to explain the embedding influence of modifying transform coefficients in I-frames and P-frames, and based on the above analyzation, a novel cost function is proposed. Next, we explore the modification of non-zero coefficients of $4 \times 4$ TUs in P-frames, which improves embedding capacity significantly while maintaining relatively low visual degradation. Last, a new indicator is introduced to measure the visual degradation under different embedding capacity. Extensive experimental results show that the proposed method outperforms in visual quality, embedding capacity and anti-steganalysis performance, compared with the existing transform coefficient-based and PU-based HEVC steganography. In future work, the non-additive cost or asymmetric cost will be further explored to improve the performance of video steganography.

**Acknowledgements.** This work was supported by the National Natural Science Foundation of China (62071267, 61771270, 62171244), Zhejiang Provincial Natural Science Foundation of China (LR20F020001).

# References

1. Wang, Y., Cao, Y., Zhao, X., Xu, Z., Zhu, M.: Maintaining rate-distortion optimization for IPM-based video steganography by constructing isolated channels in HEVC. In: 6th ACM Workshop on Information Hiding and Multimedia Security (IH and MMSec), Innsbruck, Austria, pp. 97–107. ACM (2018)
2. Jia, X., Wang, J., Liu, Y., Kang, X., Shi, Y.: A layered embedding-based scheme to cope with intra-frame distortion drift in IPM-based HEVC steganography. In: 2021 IEEE International Conference on Acoustics, Speech and Signal Processing (ICASSP), Toronto, Canada, pp. 2720–2724. IEEE (2021)
3. Dong, Y., Sun, T., Jiang, X.: A high capacity HEVC steganographic algorithm using intra prediction modes in multi-sized prediction blocks. In: Yoo, C.D., Shi, Y.-Q., Kim, H.J., Piva, A., Kim, G. (eds.) IWDW 2018. LNCS, vol. 11378, pp. 233–247. Springer, Cham (2019). https://doi.org/10.1007/978-3-030-11389-6_18
4. Dong, Y., Jiang, X., Li, Z., Sun, T., Zhang, Z.: Multi-channel HEVC steganography by minimizing IPM steganographic distortions. IEEE Trans. Multimed. https://doi.org/10.1109/TMM.2022.3150180

5. Yang, J., Li, S.: An efficient information hiding method based on motion vector space encoding for HEVC. Multimed. Tools Appl. **77**(10), 11979–12001 (2017). https://doi.org/10.1007/s11042-017-4844-1

6. Yao, Y., Zhang, W., Yu, N., Zhao, X.: Defining embedding distortion for motion vector-based video steganography. Multimed. Tools Appl. **74**, 11163–11186 (2015)

7. Guo, M., Sun, T., Jiang, X., Dong, Y., Xu, K.: A motion vector-based steganographic algorithm for HEVC with MTB mapping strategy. In: Wang, H., Zhao, X., Shi, Y., Kim, H.J., Piva, A. (eds.) IWDW 2019. LNCS, vol. 12022, pp. 293–306. Springer, Cham (2020). https://doi.org/10.1007/978-3-030-43575-2_25

8. Liu, S., Liu, B., Hu, Y., Zhao, X.: Non-degraded adaptive HEVC steganography by advanced motion vector prediction. IEEE Signal Process. Lett. **28**, 1843–1847 (2021)

9. Ma, X., Li, Z., Tu, H., Zhang, B.: A data hiding algorithm for H.264/AVC video streams without intra-frame distortion drift. IEEE Trans. Circuits Syst. Video Technol. **20**(4), 1320–1330 (2010)

10. Xue, Y., Zhou, J., Zeng, H., Zhong, P., Wen, J.: An adaptive steganographic scheme for H.264/AVC video with distortion optimization. Signal Process.: Image Commun. **76**, 22–30 (2019)

11. Chen, Y., Wang, H., Wu, H., Wu, Z., Li, T., Malik, A.: Adaptive video data hiding through cost assignment and STCs. IEEE Trans. Depend. Secure Comput. **18**(3), 1320–1335 (2021)

12. Chen, Y., Wang, H., Choo, K., He, P., Salcic, Z., Kaafar, M., et al.: DDCA: a distortion drift-based cost assignment method for adaptive video steganography in the transform domain. IEEE Trans. Depend. Secure Comput. **19**(4), 2405–2420 (2021)

13. Liu, Y., Liu, S., Zhao, H., Liu, S.: A new data hiding method for H.265/HEVC video streams without intra-frame distortion drift. Multimed. Tools Appl. **78**, 6459–6486 (2019)

14. Chang, P., Chung, K., Chen, J., Lin, C., Lin, T.: A DCT/DST-based error propagation-free data hiding algorithm for HEVC intra-coded frames. J. Vis. Commun. Image Represent. **25**(2), 239–253 (2014)

15. Zhou, A., Jiang, X., Sun, T., Li, Z., Dong, Y.: A HEVC steganography algorithm based on DCT/DST coefficients with BLB distortion model. In: 14th International Congress on Image and Signal Processing, BioMedical Engineering and Informatics (CISP-BMEI), Shanghai, China, pp. 1–9. IEEE (2021)

16. Yang, Y., Li, Z., Xie, W., Zhang, Z.: High capacity and multilevel information hiding algorithm based on PU partition modes for HEVC videos. Multimed. Tools Appl. **78**(7), 8423–8446 (2018)

17. He, S., Xu, D., Yang, L., Liu, Y.: HEVC video information hiding scheme based on adaptive double-layer embedding strategy. J. Vis. Commun. Image Represent. **87**, 103549 (2022)

18. Filler, T., Judas, J., Fridrich, J.: Minimizing additive distortion in steganography using syndrome-trellis codes. IEEE Trans. Inf. Forensics Secur. **6**(3), 920–935 (2011)

19. Zhai, L., Wang, L., Ren, Y.: Universal detection of video steganography in multiple domains based on the consistency of motion vectors. IEEE Trans. Inf. Forensics Secur. **15**, 1762–1777 (2020)

20. Wang, Y., Cao, Y., Zhao, X.: Video steganalysis based on centralized error detection in spatial domain. Infor. Secur. Cryptol. 472–483 (2017)

# Forensics and Security Analysis

# SE-ResNet56: Robust Network Model for Deepfake Detection

Xiaofeng Wang[✉], Zekun Zhao, Chi Zhang, Ningning Bai, and Xingfu Hu

Xi'an University of Technology, Xi'an, China
Xfwang66@sina.com.cn

**Abstract.** In recent years, high quality deepfake face images generated by Generative Adversarial Networks (GAN) technology have caused serious negative impacts in many fields. Traditional image forensics methods are unable to deal with deepfake that relies on powerful artificial intelligence technology. Most of the emerging deep learning-based deepfake detection methods have the problems of complex models and weak robustness. In this study, to reduce the number of network parameters, improve the detection accuracy and solve the problem of weak robustness of the detection algorithm, we propose a new lightweight network model SE-ResNet56 to detect fake face images generated by GAN. The proposed algorithm has high detection accuracy, strong robustness to content-preserving operations and geometric distortions, and strong generalization ability to different types of deepfake images generated by the same GAN.

**Keywords:** Deepfake · GAN · Lightweight network · Strong robustness

## 1 Introduction

As a product of the artificial intelligence technology, deepfake has emerged and developed rapidly in recent years. The emergence of deepfake has brought confusion to many industries, for example, criminals use deepfake to commit economic crimes through fraud, maliciously tamper video or image content to damage the appearance of public figures, and so on. Once deepfake technology is abused, its negative impact will be very serious. How to develop powerful technologies to deal with the challenges caused by deepfake has become the focus issue that is concerned by academic and industrial community.

Relying on powerful functions such as GAN [1], deepfake can generate imagery with the same distribution as the real one, so that not only human vision cannot distinguish the authenticity, also, existing detection algorithms are unable to do so. At present, the main types of face imagery deepfake include entire face synthesis, faceswap, expression transfer and attribute manipulation, etc. Among them, entire face synthesis is the most common deep forgery method. It refers to take noise as the input and uses GAN to generate high-resolution face images that do not exist in real world, as shown in Fig. 1. In fact, since the emergence of deepfake, scholars have conducted research on deepfake detection methods, and many excellent detection methods have emerged. The

X. Zhao et al. (Eds.): IWDW 2022, LNCS 13825, pp. 37–52, 2023.
https://doi.org/10.1007/978-3-031-25115-3_3

early methods are mainly based on the face physiological features and specific forgery effects introduced by deepfake. These methods have achieved high detection accuracy on some deepfake datasets. However, with the development of GAN technology, such methods cannot detect new types of deepfake, as a result, the detection performance is unsatisfactory.

**Fig. 1.** Real face and entire face synthesis

In recent years, with the development of deepfake technology, many high quality deepfake image/video datasets have emerged, which promote the rapid development of data-driven deepfake detection methods. This type of approach does not rely on any artifacts and directly trains the network model on deepfake datasets. Although data-driven detection methods have high detection accuracy, such methods have huge parameters and weak robustness. Therefore, how to construct effective network models that are tailored to suit the task of deepfake detection has become a research focus in recent years.

In this study, we aim to reduce the number of network parameters, improve the detection accuracy, and solve the problem of weak robustness of the detection algorithm, and propose a new network model to detect entire face synthesis images. The main contributions are as follows: (1) we propose a lightweight network model SE-Resnet56, this model can be used to detect high-resolution deepfake face images. (2) The proposed algorithm is provided with higher detection accuracy and strong robustness to against content-preserving manipulations and geometric distortion, and has strong generalization to different types of deepfake generated by the same GAN.

## 2   Related Works

In recent years, face images generated by GAN are so realistic that they are indistinguishable from the real images. To prevent the adverse effects caused by this technology, the detection technology against deepfake has become one of the research hotspots in the field of digital forensics and artificial intelligence. In this section, we discuss and analyze existing classical detection methods from three aspects, physiological features-based methods, specific artifact-based methods and data-driven methods.

### 2.1   Physiological Features Based Deepfake Detection Method

This type of methods are mainly aimed at deepfake videos. Since fake videos generated by AI cannot completely simulate human physiological performance, the inconsistency of physiological features can be used as important evidence for identifying deepfake videos. Y. Yang [2] observed that the 3D head posture errors would be introduced inevitably in the process of stitching fake faces into face regions in source video. The authors used the head posture difference as feature vector to train an SVM classifier to detect deepfake videos. Li [3] extracted the eye sequence of the face image and input it into the LRCN (Long-term Recurrent Convolutional Network) to obtain the state probability of the eye being open or closed, and used the state information to identify the authenticity of the video. Korshunov et al. [4] extracted the inconsistency between lip movement and speech when people speak in videos, then used feature processing classifiers and LSTM (Long Short-Term Memory) to detect fake videos. The authors of the literature [5] believed that human facial expressions and head movements would show unique patterns during speech, which is called a soft biometric model. They used Pearson correlation coefficient to measure the similarity of feature vectors, and used SVM to distinguish the true and false videos. Deepfake detection methods based on physiological features are all using the imperfection of deepfake technology. With the development of AI technology, the deepfake technology is becoming more and more sophisticated, and this type of detection methods is not persistent and universal.

## 2.2 Specific Artifact Based Deepfake Detection Method

The researchers found that visual deepfake products carry specific artifacts introduced by GAN. Although human eyes cannot perception these artifacts, machine learning methods can expose them. Mo [6] proposed a CNN-based method that detect the specific artifacts introduced by deepfake via using the residual image after high-pass filter, and achieved remarkable results.

Nirkin [7] believed that the feature information of the manipulated region in the image is different from that of the external environment, and take the consistency of two type of features as the evidence to distinguish deepfake images from real images. Li [8] proposed a new deepfake detection evidence: Face X-ray, which uses the image differences existing in the mixed boundaries of deepfake face as features to identify the authenticity of the image. Materm [9] used a series of artifacts such as inconsistent eye colors and inaccurate facial features, and used the KNN algorithm to classify deepfake images from real images. Guo [10] proposed an image preprocessing method with residual network AREN to deepen the detection of image artifacts. Chen [11] proposed a two-step framework that contains a mask-based detector and reconstructor. The detection module defines some criteria for determining anomalies, and then uses them to guide the reconstruction module to perform a learnable reconstruction process. This method has higher detection accuracy. The images generated by GAN contain specific GAN fingerprints [12], such fingerprints can be regard as the effective evidence for deepfake detection. Zhang et al. [13] proposed a spectrum-based classifier model that can achieve better detection for deepfake images generated by popular GAN models such as CycleGAN. Wang et al. [14] proposed a deepfake image detection algorithm based on monitoring the behavior of neurons. McCloskey [15] found that the images generated by GAN are similar to real images, but the spectral sensitivity is different from that of the real images. They proved that this clue is effective evidence for deepfake detection.

Although above deepfake detection methods are simple and efficient, they are only suitable for the case where the forged images contain specific GAN fingerprint artifacts. With the continuous progress of GAN technology, recent deepfake image generation models can circumvent such detection models by using fingerprint-free GANs. Therefore, these methods will be invalid. Moreover, generally, the robustness and generalization of these methods are not strong.

### 2.3  Data-Driven Deepfake Detection Method

With the development of deepfake technology, many high-quality deepfake image/video datasets have emerged, which promote the development of data-driven based deepfake detection technology. This kind of methods train the neural network as a universal classifier, and detect deepfake images through feature extraction and classification.

Zhang et al. [16] utilized SURF and bag-of-words model to extract a compact set of image features and fed them into classifiers such as SVM for detecting face-swapped images. Do et al. [17] used the fake images generated by GAN to train the classical convolutional neural network model VGG-Net to achieve the purpose of distinguishing fake images from real images. Afchar et al. [18] proposed two network architectures, Meso-4 and MesoInception-4, which showed excellent performance in detecting Face2Face and deepfake images. Zhou et al. [19] proposed a deepfake detection method that is based on a two-stream network, in which, authors used tampering artifacts and residual noise as the features for judging the authenticity of images. Hsu et al. [20] proposed a deepfake discriminator, in which, authors used a contrastive loss function to search the common features of images generated by different GANs and detected deepfake images by cascading classifiers. Zhao [21] proposed a network structure based on spatial attention mechanism, and introduced a new region independence loss to solve the problems such as network learning difficulties. The above methods have achieved high detection accuracy in deepfake datasets. However, the detection efficiency is not satisfactory because most of network models have huge amount of parameters.

## 3  The Proposed Method

In this section, we describe the proposed method in detail. The proposed method including improved residual block, network constructing and deepfake detection.

### 3.1  Improved Residual Block SE-Res-Block

Considering that the inter-class gap between deepfake images and digital photos is far less than that of the image classification tasks of the pattern recognition field, therefore, data-driven-based deepfake detection method requires that the network model have excellent feature extraction abilities, so that the subtle and specific features introduced by deepfake can be extracted. To this end, we improve the resnet18 [22] and design a new network model SE-Resnet56. Our improvement aims at two aspects.

(1) To extract high-level semantic features, the depth of the network model must be concerned. However, for standard CNN, with the deepening of network layers, some problems such as gradient disappearance and performance degradation will occur, these problems will cause that network model is increasingly difficult to train. Simultaneously, the deepening of network layers will lead to a sharp increase in the number of parameters, which will increase the complexity of the network model and cause the problem of overfitting. To solve these problems, we improve the residual structure of resnet18 [22]. We introduce a bottleneck structure to establish a lightweight residual network called Resnet56. The bottleneck structure can reduce the number of the network parameters without sacrificing the feature extraction ability of the network model, thus it can accelerate the training process of the network model.

(2) The deepfake detection task is very different from the image classification task in the field of pattern recognition. The latter focuses on the difference of the global features, while the former focuses on the local features of the facial region. To strengthen the local region attention capability of the network model, we introduce SE-Net [23] into our network framework to construct a lightweight attention residual network SE-Resnet56. Our improvement process is shown in Fig. 2. Figure 2(a)

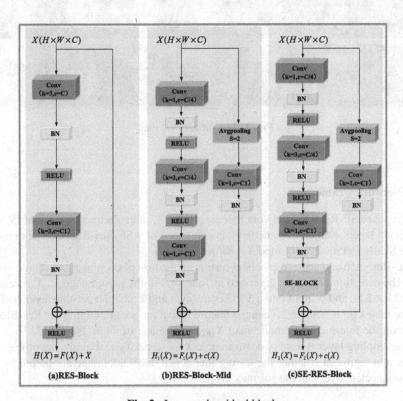

**Fig. 2.** Improved residual block

is the residual structure of the original Resnet18. Figure 2(b) is the improved residual structure via introducing the bottleneck structure. Figure 2(c) is the SE-Resnet56 constructed via introducing the attention module.

The proposed SE-Resnet56 can extract subtle specific features introduced by deepfake. This is because the original residual structure (as shown in Fig. 2(a)) pays more attention to the global feature, however, the global feature does not play the key role in the detection task of the deepfake since the images generated by GAN have the same statistical distribution as the camera photos, while the proposed SE-Resnet56 pays more attention to the local features of the facial region.

We compared the heat map of the network output, as shown in Fig. 3. In Fig. 3, the first row represents the heat map obtained using the structure of Fig. 2(c), and the second row represents the heat map obtained using the structure of Fig. 2(b). As can be seen from Fig. 3, SE-Block pays more attention to the facial features of the face image, this is the main reason why SE-Resnet56 has higher detection accuracy.

**Fig. 3.** Face image heat map

## 3.2 Network Framework

The proposed SE-Resnet56 network model includes 37 convolutional layers and 19 fully connected layers, and the core network is composed of 9 SE-Res-Blocks. The diagram of the SE-Resnet56 network model is shown in Fig. 4.

SE-Resnet56 can achieve excellent results on the task of deepfake face image detection. The detection process is described as follows. For the input image $X$, $X$ is resized into $224 \times 224$, and is input into a $3 \times 3$ conv layer and $2 \times 2$ maxpooling layer, and the feature map $X_{MP}$ is obtained. Then the feature map $X_{MP}$ goes through 9 SE-Res-Blocks to obtain the intermediate feature map $X_{SRB}$ with shape of $14 \times 14 \times 512$. Then the average-pooling layer is used to down sample $X_{SRB}$ to get $X_{AVG}$. Finally, the probability values that represent whether the testing image is a deepfake image are output by the fully-connected layer and the softmax layer.

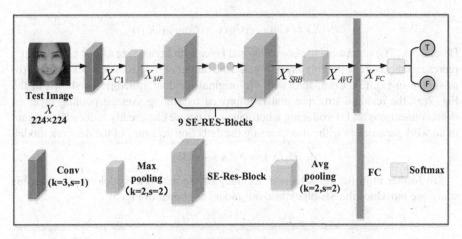

**Fig. 4.** The diagram of the SE-Resnet56 network model

## 3.3 Deepfake Face Image Detection Algorithm Based on SE-RESNET56

IN this section, we establish a deepfake image detection algorithm based on SE-RESNET56, which includes three main phases: image preprocessing, deepfake saliency feature extraction and deepfake detection.

### 3.3.1 Image Preprocessing

To strengthen the robustness of the algorithm, we adopted geometric enhancement strategies such as random cropping, random horizontal flipping, and random offset in vertical and horizontal directions for the input image. At the same time, to speed up the convergence speed of the network model, the pixel values of the image are normalized.

### 3.3.2 Deepfake Saliency Feature Extraction

To extract deepfake saliency feature, the input image is fed into the backbone network. The core component of the backbone network is the SE-Res-block, as shown in the Fig. 2(c), it is mainly composed of three parts: residual structure, bottleneck structure and channel attention module.

Different from the residual structure of Resnet18, we introduce an average-pooling layer and a $1 \times 1$ convolutional layer into the short connection to adjust the size of the feature map so that the output of the short connection can match the subsequent linear superposition, as shown in Eq. (1). Meanwhile, we introduce a bottleneck structure, in which, a $1 \times 1$ convolution layer is used to reduce the dimension of the input feature map, and the channel number of the feature map is compressed to 1/4. Then we use a $3 \times 3$ convolutional layer for feature extraction, and finally, we use a $1 \times 1$ convolutional layer to output the feature map with the specified number of channels according to the needs of the network, as shown in Eq. (2).

$$c(X) = Cov_{1 \times 1}(Avg(X)) \tag{1}$$

$$F_1(X) = Cov_{1\times1}(Cov_{3\times3}(Cov_{1\times1}(X)))  \tag{2}$$

Here, $Cov_{k\times k}(\cdot)$ represents the convolutional layer with kernal size of $k \times k$, and $Avg(\cdot)$ represents the average pooling layer. Finally, the output of the residual block in Fig. 2(b) as shown in Eq. (3). Compared with the original residual structure, as shown in the Fig. 2(a), the residual structure that is improved by adding average pooling layer on short connections and introducing a bottleneck structure can greatly reduce the amount of network parameters without decreasing the detection accuracy of the network model.

$$H_1(X) = F_1(X) + c(X)  \tag{3}$$

To further improve the detection accuracy of the network model proposed in this study, we introduce the SE-block into our model as shown in Fig. 5.

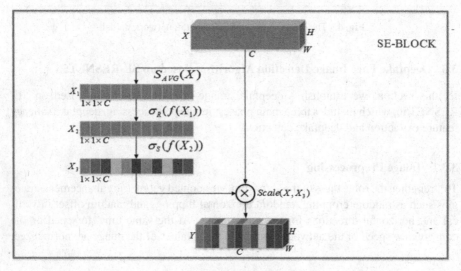

**Fig. 5.** SE-Block

SE-Block (Squeeze-and-Excitation Block) is a kind of channel attention mechanism. In order to recalibrate the features of each channel, SE-Block makes use of the dependent relationship between different channels in the feature map to assign different weights to each channel. SE-block mainly consists of two stages, the squeeze stage and the excitation stage. To fully extract the global information of the image, the global average-pooling process is performed on each channel of the feature map in the squeeze stage, as shown in Eq. (4).

$$X_1 = S_{AVG}(X) = \underset{a=1}{\overset{C}{Contact}}(\frac{1}{H \times W} \sum_{i=1}^{W} \sum_{j=1}^{H} x_a(i,j))  \tag{4}$$

Here, $C$ represents the number of channel of the feature map, $x_a(i,j)$ represents the value at $(i,j)$ of $a$-th channel, $H$ and $W$ represent the width and height of the feature map, respectively.

In the excitation stage, the fully-connected layer is used to obtain the information dependency between channels, and then reasoned out different weights of the feature information in each channel. The calculation process is shown in Eq. (5) and Eq. (6).

$$X_2 = \sigma_R(f(X_1)) \tag{5}$$

$$X_3 = \sigma_S(f(X_2)) \tag{6}$$

$f(\cdot)$ represents the fully connected layer, $\sigma_R$ and $\sigma_s$ represent the Relu and Sigmoid activation functions, respectively. By multiplying the weight vector $X_3$ and the original inputs $X$ channel by channel, the final output $Y$ can be obtained, as shown in Eq. (7).

$$Y = SE(X) = X_1 \otimes X_3 \tag{7}$$

Finally, as shown in Fig. 2c, the output of SE-RES-Blcok is $H_2(X)$.

$$H_2(X) = SE(F_1(X)) + c(X) \tag{8}$$

Here, $c(X)$ represents the output of the improved short connection.

### 3.3.3 Deepfake Detection

The detection stage consists of a fully connected layer and a softmax layer. The function of the fully-connected layer is to map the distributed features to the sample label space, and the softmax function maps the output of the fully connected layer to probability values, the computing process is shown in Eqs. (9) and (10).

$$Z_i = e^{r_i}, i = 1, 2 \tag{9}$$

$$P_i = \frac{z_i}{Z_1 + Z_2}, i = 1, 2 \tag{10}$$

Here, $r_i$ and $P_i$ represent the output of the fully connected layer and the output of the softmax layer, respectively.

The loss function is used to measure the gap between the predicted data and the real data. In the training process of our network model, we use the cross-entropy loss function as the model training feedback signal. The mathematical expression is shown in Eq. (11), where $y_i$ represent the label values.

$$L = -\sum_{i=1}^{2} y_i \log P_i \tag{11}$$

## 4 Experimental Results and Performance Analysis

In this section, we evaluate the performance of the proposed SE-Resnet56 through experiments. The experiments are carried out on a computer with Win10 system, i7-6800K CPU GTX 1080Ti GPU, 32.00 GB RAM. The training process and detection process are implemented on the Keras deep learning platform, using version 1.13. The hyperparameters used in the experiments are shown in Table 1.

| Imagesize | Channels | Learning rate | Epoch | Batch size | Step |
|-----------|----------|---------------|-------|------------|------|
| 224 × 224 | 3 | 0.001 | 100 | 32 | 600 |

### 4.1 Evaluation Indicators and Dataset Settings

In the experiment, we assume real face images as positive samples and forged face images as negative samples. TP represents the number of practical positive samples that are classified as positive samples; TN represents the number of practical negative samples that are classified as negative samples; FN: represents the number of practical positive samples that are classified as negative samples; FP represents the number of practical negative samples that are classified as positive samples. We used the following evaluation indicators: Accuracy, Precision, Recall, F1-measure and Params. They are defined as follows:

$$\text{Accuracy} = \frac{TP + TN}{TP + TN + FN + FP} \tag{12}$$

$$\text{Precision} = \frac{TP}{TP + FP} \tag{13}$$

$$Recall = \frac{TP}{TP + FN} \tag{14}$$

$$F_1 = \frac{(1 + \beta^2) \cdot Recall \cdot Precision}{\beta^2 \cdot Recall + Precision}; \beta = 1 \tag{15}$$

To test the learning ability of the SE-Resnet56 network model, we generate experimental datasets by sampling partial deepfake images from StyleGAN and StyleGAN2 in the deepfake database and CelebA. We randomly sample 4500 fake face images from StyleGAN2's star face dataset, Internet celebrity face dataset, and yellow face dataset, respectively. We randomly sample 4500 fake face images from StyleGAN's adult face dataset, randomly sample 4500 fake face images from the CelebA dataset, as well as the real face dataset corresponding to each category. Then we build our experimental dataset that contains 45000 images.We construct the training set, test set and validation set with a ratio of 4:4:1, and there are no cross-used images in each dataset.

### 4.2 Ablation Experiment

To illustrate the importance of bottleneck structures and attention mechanisms for deep forgery detection, we perform the ablation experiments. In the experiments, model-NB represents the network model that remove bottleneck structure from SE-Resnet56, model-NS represents the network model that remove SE-Block from SE-Resnet56. The experimental results are shown in Table 2.

As can be seen from Table 2, the introduction of the bottleneck structure in the residual block reduces the amount of model parameters, which effectively reduces the

**Table 2.** The result of the ablation experiment

|             | Accuracy | Precision | F1-measure | Params  |
|-------------|----------|-----------|------------|---------|
| Model-NB    | 0.9670   | 0.9499    | 0.9676     | 8237246 |
| Model-NS    | 0.9265   | 0.8662    | 0.9197     | 1292962 |
| SE-Resnet56 | 0.9851   | 0.9778    | 0.9853     | 1382622 |

computational complexity of the model, and alleviates the risk of overfitting. Further-more, the introduction of SE-block improves the detection accuracy of the network model. The reason is that the SE-Block can use the correlation between channels to extract more detailed features, so as to force our network model to pay more attention to the local features of the facial region. To visually illustrate that SE-block has a strong ability to extract detailed features, we visualize the output feature map of the deepfake image through three residual blocks of SE-Resnet56 and Model-NS, respectively. The results are shown in Fig. 6.

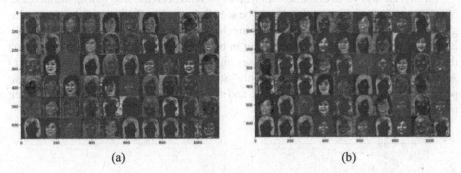

(a)                                             (b)

**Fig. 6.** Feature layer visualization. (a) The output of model-NS; (b) The output of SE-Resnet56

As can be seen from Fig. 6, the number of the activated feature maps of SE-Resnet56 is more than that of model-NS. This illustrates that the SE-Block can use the correlation between channels to extract more effective features for deepfake detection, so as to improve the detection performance of network models.

To further demonstrate the importance of SE-Block for deepfake detection, we compare the visual effect of the same channel feature maps that are extracted by the first residual block of SE-Resnet56 and model-NS, respectively, as shown in Fig. 7.

As can be seen from Fig. 7(a), the feature maps generated by the model-NS mainly contain texture and color information rather than content information, and it is difficult to distinguish the content of the face and the edge from the background. However, the feature maps extracted by the SE-Resnet56 clearly show the outline and edge of the facial regions, and the detailed features of the facial information that are of great significance for deepfake detection.

Fig. 7. The visual effect of the feature maps. (a) Feature maps for the $8^{th}$, $16^{th}$, $24^{th}$ channels extracted by the SE-Resnet56, (b) Feature maps for the $8^{th}$, $16^{th}$, $24^{th}$ channels extracted by the model-NS.

## 4.3 The Performance Analysis

In this section, we train the SE-Resnet56 network model by using mini-batch gradient descent and optimize the network by using the Adam optimizer. Figure 8 shows the loss and accuracy on the training set and validation set during the training process. At the same time, in the case that the detection accuracy does not increase after 15 consecutive epochs in the validation set, we drop the learning rate to one-tenth of the original learning rate.

Ideally, the loss value of the detection model should decrease slowly with the increase of the number of training epochs. As can be seen from the Fig. 8, the loss value fluctuates greatly at the initial stage of model training, however, with the increase of epochs, the loss value fluctuates within a small range in both training set and validation set. Also, the detection accuracy gradually increases on both the training set and validation set. Additionally, we adopt the "Save best only" strategy in model training to keep the best performance during the iteration process, and avoid the influence of loss shock on model training.

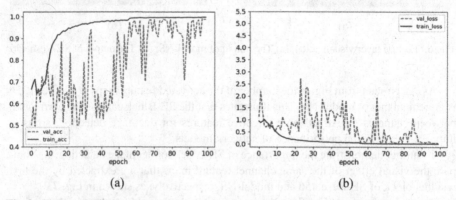

Fig. 8. The loss and accuracy on the training dataset and validation dataset during the training process, (a) The detection accuracy, (b) The loss value

We compare the detection performance of the SE-Resnet56 with other algorithms, the results are shown in the Table 3.

As can be seen from Table 3, Accuracy, Recall, F1 and AUC of the proposed SE-Resnet56 are all superior to the comparative models. This means that the proposed

**Table 3.** The comparison of the detection performance

|  | Accuracy | Recall | F1-measure | AUC | Params |
|---|---|---|---|---|---|
| Meso4 [18] | 0.7157 | 0.4350 | 0.6047 | 0.9702 | 24250 |
| Inception [24] | 0.9344 | 0.9263 | 0.9204 | 0.9901 | 6739730 |
| Resnet-18 [22] | 0.9667 | 0.9905 | 0.9675 | 0.9964 | 12568194 |
| Resnet-50 [22] | 0.9125 | 0.9815 | 0.9179 | 0.9839 | 23565250 |
| SE-Resnet56 | 0.9851 | 0.9925 | 0.9853 | 0.9987 | 1382622 |

SE-Resnet56 is provided with excellent performance for deepfake detection. Moreover, the network model is lightweight, and the number of parameters is relatively small. Although the parameter quantity of our model is slightly more than that of the Meso4, the detection accuracy is improved more than 25% over the Meso4 model. The index value is significantly reduced, down by an order of magnitude.

To further analyze the reason why our model is superior to Meso4, we visualize the class activation map of Meso4 versus SE-Resnet56, as shown in Fig. 9. In Fig. 9, the first row represents the class activation heatmap of SE-Resnet56, and the second row represents the class activation heatmap of Meso4.

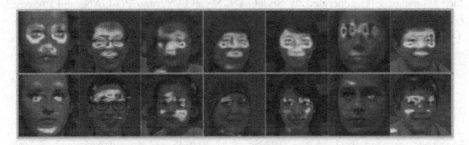

**Fig. 9.** Heatmap of Meso4 and SE-RESNET56 models

It can be seen from Fig. 9 that for deepfake face images, Meso4 cannot extract high-level semantic features, such as facial features, while the SE-Resnet56 model customized for deepfake detection tasks has a strong ability to extract the detailed information of the facial features that are very beneficial for deepfake detection.

## 4.4 Robustness Analysis

Robustness is an important indicator for evaluating the performance of a detection algorithm. To test the robustness of the proposed model, we perform content-preserving manipulations such as adding salt-pepper noise (noise factor = 0.0005), gamma correction and Gaussian blurring on the images in the test set, and use ResNet18, Inception and SE-Resnet56 to detect these manipulated images. The test results of the detection accuracy are shown in Table 4.

**Table 4.** The detection accuracy under gamma correction, Gaussian blurring and histogram equalization

| Operation | No-ops | GB | GC | SP-Noise | HE |
|---|---|---|---|---|---|
| Resnet18 [22] | 0.9667 | 0.9414 | 0.9278 | 0.9380 | 0.9337 |
| Inception [24] | 0.9344 | 0.8920 | 0.9022 | 0.9177 | 0.8940 |
| Resnet50 [22] | 0.9125 | 0.8937 | 0.9003 | 0.8763 | 0.8874 |
| SE-Resnet56 | 0.9850 | 0.9736 | 0.9682 | 0.9774 | 0.9652 |

In the experiments shown in Table 4, the images in test set are manipulated by Gaussian Blur (GB), Gamma Correction (GC), Salt-Pepper noise (SP-Noise) and Histogram Equalization (HE), respectively. It can be seen from the experimental results that the detection accuracy of the proposed model is satisfactory. The detection performance of the proposed method consistently outperforms that of the contrastive models. This indicates that SE-Resnet56 is provided with satisfactory robustness.

### 4.5  Generalization Analysis

To investigate the generalization of the SE-Resnet56, our test datasets uses images (Entire face synthesis, Efs) generated by StyleGAN, StyleGAN2 and PG-GAN, and images (Attribute manipulation, Amp) generated by StyleGAN. The test results are shown in Table 5.

**Table 5.**  Model generalization

| Testdata | StyleGAN | | PGGAN | StyleGAN2 |
|---|---|---|---|---|
| | Efs | Amp | Efs | Efs |
| Accuracy | 0.9832 | 0.9795 | 0.9845 | 0.9865 |
| Recall | 0.9930 | 0.9943 | 0.9925 | 0.9885 |
| Precision | 0.9803 | 0.9793 | 0.9836 | 0.9855 |

Here, when we test with the new data generated by the original GAN (The training set include forged face images generated by this kind of GAN), the detection accuracy of SE-Resnet56 does not drop significantly, and for another deepfake type (Attribute manipulation), the detection performance is still satisfactory.

## 5  Conclusion

In this study, we introduce SE-Block and bottleneck structure into resnet18 and present a lightweight network model SE-ResNet56. The proposed method can be used to detect high-resolution deepfake face images. Compared with other advanced network models,

SE-ResNet56 shows better detection performance and stronger robustness for deepfake images, and it can also show excellent detection performance for another deepfake type (attribute manipulation) generated by the same GAN.

**Acknowledgments.** Many thanks to the anonymous reviewers for their valuable comments to improve our work. This work was supported by the National Natural Science Foundation of China under Grant No. 61772416; Shaanxi province key research and development project, No. 2022GY-087.

# References

1. Karras, T., Aila, T.: Progressive growing of GANs for improved quality, stability, and variation (2017)
2. Yang, X., Li, Y.: Exposing deep fakes using inconsistent head poses. In: International Conference on Acoustics, Speech and Signal Processing (ICASSP), Brighton, UK, pp. 8261–8265 (2019)
3. Li, Y., Chang, M.C.: In ICTU Oculi: exposing AI generated fake face videos by detecting eye blinking. In: IEEE International Workshop on Information Forensics and Security (WIFS), pp. 1–7 (2018)
4. Donahue, J., Hendricks, L.A.: Long-term recurrent convolutional networks for visual recognition and description. IEEE Trans. Pattern Anal. Mach. Intell. **39**(4), 677–691 (2017)
5. Shan, M., Tsai, T.J.: A cross-verification approach for protecting world leaders from fake and tampered audio (2020)
6. Mo, H.X., Chen, B.L.: Fake faces identification via convolutional neural network. In: Workshop on Information Hiding and Multimedia Security, Innsbruck, Austria, pp. 43–47. ACM (2018)
7. Nirkin, Y., Wolf, L.: Deepfake detection based on discrepancies between faces and their context. IEEE Trans. Pattern Anal. Mach. Intell. (2021)
8. Li, L., Bao, J.: Face X-ray for more general face forgery detection. In: IEEE/CVF Conference on Computer Vision and Pattern Recognition (CVPR), pp. 5000–5009 (2020)
9. Matern, F., Riess, C.: Exploiting visual artifacts to expose deepfakes and face manipulations. In: IEEE Winter Applications of Computer Vision Workshops (WACVW), pp. 83–92 (2019)
10. Guo, Z.Q., Yang, G.B.: Fake face detection via adaptive residuals extraction network [EB/OL] (2021). https://www.researchgate.net/publication/341310624.Pdf
11. Chen, Z.K., Xie, L.X.: Mask-guided detection and reconstruction for defending deepfakes [EB/OL] (2021). https://arxiv.org/pdf/2103.14211.Pdf
12. Marra, F., Gragnaniello, D.: Do GANs leave artificial fingerprints. In: 2019 IEEE Conference on Multimedia Information Processing and Retrieval (MIPR), pp. 506–511 (2019)
13. Zhang, X., Karaman, S.: Detecting and simulating artifacts in GAN fake images. In: IEEE International Workshop on Information Forensics and Security (WIFS) (2019)
14. Wang, R., Ma, L.: A simple yet robust baseline for spotting AI-synthesized fake faces. In: International Joint Conferences on Artificial Intelligence (IJCAI 2020), pp. 3444–3451 (2019)
15. McCloskey, S., Albright, M.: Detecting GAN-generated imagery using color cues [EB/OL] (2018). https://arxiv.org/pdf/1812.08247.Pdf
16. Zhang, Y., Zheng, L.L.: Automated face swapping and its detection. In: IEEE 2nd International Conference on Signal and Image Processing (ICSIP), pp. 15–19. IEEE (2017)
17. Do, N.T., Na, I.S.: Forensics face detection from GANs using convolutional neural network. In: International Symposium on Information Technology Convergence (ISITC), pp. 376–379 (2018)

18. Afchar, D., Nozick, V.: MesoNet: a compact facial video forgery detection network. In: IEEE International Workshop on Information Forensics and Security (WIFS), Hong Kong, China, pp. 1–7. IEEE (2018)
19. Zhou, P., Han, X.T.: Two-stream neural networks for tampered face detection. In: IEEE Conference on Computer Vision and Pattern Recognition Workshops (CVPRW), pp. 1831–1839. IEEE (2017)
20. Hsu, C., Lee, C.: Learning to detect fake face images in the wild. In: International Symposium on Computer, Consumer and Control (IS3C), pp. 388–391 (2018)
21. Zhao, H.Q., Zhou, W.B.: Multi-attentional deepfake detection [EB/OL] (2021). https://arxiv.org/pdf/2103.02406.Pdf
22. He, K., Zhang, X.: Deep residual learning for image recognition. In: IEEE Conference on Computer Vision and Pattern Recognition (CVPR), pp. 770–778 (2016)
23. Hu, J., Shen, L.: Squeeze-and-excitation networks. In: IEEE Conference on Computer Vision and Pattern Recognition (CVPR), pp. 7132–7141 (2018)
24. Szegedy, C., Liu, W.: Going deeper with convolutions. In: IEEE Conference on Computer Vision and Pattern Recognition (CVPR), pp. 1–9 (2015)

# Voice Conversion Using Learnable Similarity-Guided Masked Autoencoder

Yewei Gu[1,2], Xianfeng Zhao[1,2(✉)], Xiaowei Yi[1,2], and Junchao Xiao[1,2]

[1] State Key Laboratory of Information Security, Institute of Information Engineering, Chinese Academy of Sciences, Beijing 100195, China
{guyewei,zhaoxianfeng,yixiaowei,xiaojunchao}@iie.ac.cn
[2] School of Cyber Security, University of Chinese Academy of Sciences, Beijing 100195, China

**Abstract.** Voice conversion (VC) is an important voice forgery method that poses a serious threat to personal privacy protection, especially with remarkable achievements in timbre modification. To support forensic research on converted speech and further enrich the sources of fake speech, it is imperative to investigate new robust VC methods. VC is also considered a typical style transfer task, where style refers to speaker identity, suggesting that achieving sufficient feature decoupling is the key to obtaining robust performance. However, mainstream decoupling methods based on information-constrained bottlenecks still fail to obtain robust content-style trade-offs. In this paper, we propose a learnable similarity-guided mask (LSGM) algorithm to address the robustness problem. First, to make feature decoupling independent of specific language constructs and more applicable to diverse content, LSGM performs inter-frame feature compression only relying on the similarity of adjacent frames instead of complex inter-frame content correlation. Second, we implement feature compression by masking instead of dimensionality reduction, so no additional modules are needed to convey the speech frame length information. Moreover, we propose MAE-VC by using LSGM, which is an end-to-end masked autoencoder (MAE) with self-supervised representation learning. Experimental results indicate that MAE-VC performs comparable to state-of-the-art methods on speaker similarity and significantly improves the performance on content consistency.

**Keywords:** Voice conversion · Feature decoupling · Style transfer · Learnable similarity-guided mask · Masked autoencoder

## 1 Introduction

Voice conversion (VC) is a style transfer technology that converts the speaker style of the source speech to that of the target speech while maintaining the consistency of the content information. As an important speech forgery technique,

This work was supported by National Key Technology Research and Development Program under 2020AAA0140000.

common VC methods are not as advanced as text-to-speech (TTS) [22] or voice clone [12], but the potential challenges to social security cannot be ignored. Unfortunately, there are few detection methods dedicated to converted speech. To support the development of corresponding detection techniques, research on robust VC methods is imperative.

With the remarkable development of deep feature representation, voice conversion patterns have also undergone profound changes. To meet the training needs of deep learning, the database to support research has gradually shifted from a small amount of parallel corpus [25] to abundant non-parallel corpus [10], thereby, related research hotspots have become more colorful and interesting: from one-shot [14] to zero-shot [21]; from a single language to cross languages, *etc.* From an information-theoretic perspective, speech can be divided into two independent and complementary parts: speaker-dependent information (SDI) and speaker-independent information (SII). The common framework is to build an encoder-decoder model to decouple speech into SDI and SII via a speaker encoder and a content encoder, respectively, and then both are coupled into the decoder to predict the transformed acoustic features. To transfer arbitrary voice styles, significant efforts have been devoted to feature disentanglement. However, how to achieve sufficient decoupling of SDI and SII remains a challenge for robust conversion performance.

In terms of speech feature disentanglement, a common yet effective approach is to use information-constrained bottlenecks for compression coding. Compared with images, the 2D mel-spectrogram of speech used for conversion has a unique physical meaning, where the horizontal axis represents the frame dimension and the vertical axis represents the frequency dimension. According to whether the frame dimension is compressed, feature decoupling can be divided into two categories: utterance-level decoupling and frame-level decoupling. However, for utterance-level decoupling, frame-dimensional compression relies on inter-frame content correlation, which makes feature representation limited to specific language structures and cannot adapt to diverse content. In terms of frame-level decoupling, although it relies less on content information, it only compresses in the single frequency dimension, which is not flexible enough in feature decoupling. In addition, frame-level methods struggle to deal with redundant frames, such as noisy and silent segments, ultimately leading to undesired performance. Therefore, how to realize the complementarity of the two schemes becomes the focus of feature decoupling research.

In this paper, we propose a learnable similarity-guided mask (LSGM) algorithm to solve the fundamental robustness-flexibility dilemma. LSGM is inspired by masked autoencoder (MAE) [7], which shows that for natural signals, masked points can be inferred from surrounding similar points. Compared with common feature decoupling methods, LSGM employs a learnable network to measure the similarity of adjacent frames and then masks the most redundant frames. LSGM is considered a self-referential strategy without considering inter-frame content dependencies. Due to the short-term stationarity of speech, LSGM can significantly improve the efficiency of inter-frame compression by masking repetitive and interfering information. Furthermore, compared with artificial similarity threshold division, we use the frame mask ratio to control the number of frames to be masked, avoiding over-masking in extreme conditions. In addition, the masking strategy does not reduce the number of frames, which means that no additional module is needed to save the speech duration information. Based on LSGM, we further propose MAE-VC, an end-to-end self-supervised model based on masked autoencoder (MAE). Compared to the random masks in MAE, the model with LSGM takes full advantage of speech intrinsic characteristics and performs a more effective masking strategy. To the best of our knowledge, MAE-VC is the first method to apply learnable similarity-guided masks to speech feature decoupling in the VC domain. Results indicate the method outperforms state-of-the-art methods on content consistency while performing closely towards the best on speaker similarity.

## 2 Related Work

In the early days, due to the limitation of feature representation capability, research on VC is mainly based on a small amount of parallel corpus. Traditional statistical methods such as Gaussian mixture models (GMM) [26], frequency warping [5], and Dynamic Kernel Partial Least Squares (DKPLS) [8] are introduced to establish phonetic unit mappings between source and target speakers. There are also some non-parametric techniques such as vector quantization [30] and non-negative matrix factorization (NMF) [19]. With the development of deep learning, speech representation technology has made significant progress. To support the training of large-scale neural networks, the database has gradually shifted from parallel corpus to rich and readily available non-parallel data. According to the generative model, these methods can be divided into two categories: encoder-decoder-based models and generative adversarial network-based (GAN-based) models. The GAN-based approaches are inspired by the image style transfer framework [3,32], and then further develop the CycleGAN-VC series [15] and STARGAN-VC series [14]. In particular, VoiceMixer [17] can effectively decompose and transfer speech styles through similarity-based information bottlenecks and adversarial feedback.

**Feature Decoupling.** In terms of encoder-decoder-based models, the key to achieving robust performance lies in achieving sufficient feature decoupling. Based on variational autoencoders (VAE) [16], VAE-VC [9] employs a well-designed encoder to learn speaker-independent representations and a specifical decoder to reconstruct the target style. ACVAE-VC [13] proposes an auxiliary VAE-based classifier VAE to perform many-to-many VC, which facilitates the correct prediction of attribute classes of converted speeches. Inspired by AdaIN [11], ADAINVC [4] first applies instance normalization (IN) [27] into the VC domain, which suggests that cascaded INs in the content encoder can effectively improve the decoupling of content information. MediumVC [6] splits the conversion into a two-step process, where the source speech is first transferred

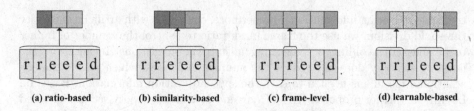

**Fig. 1.** Four feature decoupling methods based on information-constrained bottlenecks. Matches "r", "e" and "d" for frame with different content. The arc represents the frame range within a single process. The arrow points to single processing results.

into a preset middleman speech and then the middle speech combines the target speaker style into converted speech. Moreover, AutoVC [21] introduces the speaker verification system (SV) [29] to characterize speaker identities. FragmentVC [18] introduces Wav2Vec2 [1] to represent content information, and uses attention mechanism to model the mapping of mel-spectrums. Based on Unet [23], AGAINVC [2] applies activation guidance and adaptive instance normalization (AGAIN) to improve content-style trade-offs and appropriate activations are used as informative bottlenecks for content embeddings.

## 3 Backgroud

### 3.1 Feature Decoupling for SII

For time-invariant SDI (style), it is relatively stable to decouple because it is considered to be evenly distributed across frames. Therefore, in this section, present mainstream decoupling methods for SII are investigated, which are also the core of content encoders.

These methods can be further divided into 4 categories: ratio-based decoupling [4], similarity-based decoupling [17], frame-level decoupling [2,6], and learnable-based decoupling [17]. We use $X$ to represent the input feature map with the frequency dimension $D$ and frame dimension $L$. As shown in Fig. 1, the first two belong to utterance-level decoupling, compressing $X \in \mathbb{R}^{D \times L}$ into $X_c \in \mathbb{R}^{D_c \times L_c}$, where $L_c < L$. For ratio-based decoupling, where $L_c = L * \alpha$ and $\alpha$ is the preset ratio, inter-frame semantic associations are deeply explored and compressed. However, an ideal robust model should be content-agnostic, which means over-reliance on semantics will weaken the adaptability of the decoupling modules to diverse content. Instead of it, the similarity-based method, employing a self-referential decoupling strategy, only focus on inter-frame similarity rather than semantic information, where consecutive redundant or near-duplicate frames are aggregated into a single frame. However, the similarity division threshold based on traditional statistical experience makes $L_c$ unpredictable and difficult to adapt to changing speech. Often too small a threshold can lead to over-compression, and a too large one will greatly reduce the efficiency of decoupling. Also, since $L_c$ is uncontrollable, an extra module is needed

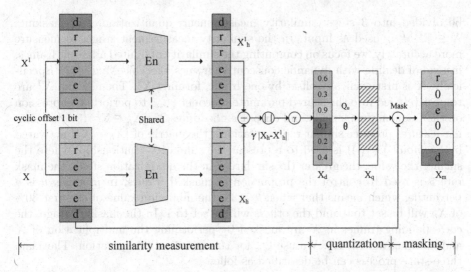

**Fig. 2.** The workflow of LSGM. In the similarity measurement stage, the similarity matrix $X_d$ of adjacent frames is calculated. During the quantization stage, $X_d$ is converted to a mask vector $X_q$ containing only 0s and 1s. In the masking stage, input frames corresponding to 0 values in $X_q$ are masked.

to deliver the number of frames. In contrast, the frame-level method accomplishes the independent decoupling of each frame by sharing parameter units without considering inter-frame relationships, while maintaining speech duration ($L_c = L$). However, the method relies only on inter-frequency compression, which is inflexible for dealing with noise and silent segments, preventing the realization of sufficient feature decoupling.

Is it possible to combine the advantages of frame-level decoupling and utterance-level decoupling to be independent of content and flexible to handle redundant information? Based on the above analysis, we propose the learnable-based decoupling, also called the learnable similarity-guided mask (LSGM) method. There are three advantages. First, similar to the similarity-based method [17], LSGM does not rely on content correlation but performs inter-frame compression based on the similarity of limited adjacent frames. Second, the method is the first to use a learnable neural network to measure inter-frame similarity instead of traditional statistical methods such as cosine similarity. Third, inter-frame compression is achieved by masking redundant frames rather than reducing the number of frames, thus maintaining the consistency of speech duration.

## 4   Method

### 4.1   LSGM

To improve the efficiency of feature compression, LSGM is used to mask redundant and noisy information. As shown in Fig. 2, the workflow of LSGM can be divided into 3 steps: similarity measurement, quantization, and masking. $X \in \mathbb{R}^{D \times L}$ is used as input. In the similarity measurement stage, to measure more accurately, we focus on computing the similarity of limited adjacent frames, instead of dealing with dynamic consecutive frames like VoiceMixer [17]. Specifically, $X$ is first cyclically offset by one frame, forming $X^1$. Then $X$ and $X^1$ are respectively fed into the shared parameter encoder ($E_n$) to perform compression on the frequency dimension and output embeddings $X_h, X_h^1 \in \mathbb{R}^{1 \times L}$. Since similar features produce similar representations, the matrix of $\left| X_h - X_h^1 \right|$ activated by $Sigmoid(\cdot)$ ($\gamma(\cdot)$) is used to represent the similarity relationship, where the smaller the value, the greater the similarity. In the quantization stage, the mask rate $\theta$ is used to control the proportion of masked frames. In practice, $\theta$ is a percentile, which means that when $\theta = 30$, the minimum value of the first 30% of $X_d$ will be set to 0, and the others will be set to 1. In the masking stage, the corresponding frames in $X$ are masked by performing the multiplication of $X$ and masking matrix $X_q$. We use $Q_u$ for the quantization operation. The main three-stage process can be described as follows.

$$X_d = \gamma(\left| E_n(X) - E_n(X^1) \right|) \tag{1}$$
$$X_q = Q_u(X_d, \theta) \tag{2}$$
$$X_m = X * X_q \tag{3}$$

**Backward Propagation of LSGM.** To make LSGM learnable, we must obtain the gradients of the weights that yield $X_d$. However, as the quantization operation ($Q_u$) is non-differentiable and prevents the propagation of gradients, the relevant parameters cannot be optimized. Therefore, the approximate gradient of $X_d$ should be designed in an alternative way, where forward and backward propagation are described in detail.

In forward propagation, quantization is denoted as Eq. (2), where $X_d \in [0, 1]$; $\theta$ is a constant, $X_q$ contains only 0 and 1. For optimization, we first introduce $\hat{X}_q$ as the substitution for $X_q$. It can be obtained by Eq. (4), where $\mathcal{M}(\cdot)$ represents the MinMax normalization function that scales the value to $[0, 1]$, and $\odot$ represents for element-wise multiplication. Due to $X_q \in [0, 1]$, after the processing of Eq. (4), the high value in $\hat{X}_q$ will be biased towards 1 and the low value will be close to 0, making $\hat{X}_q$ close to $X_q$.

$$\hat{X}_q = \mathcal{M}(\frac{1}{2}X_d \odot X_d) \tag{4}$$

In backward propagation, as $\mathcal{M}(\cdot)$ does not affect the magnitude relationship of the values, the partial derivative of $X_q$ can be approximated by Eq. (5). We use $\nabla.\mathcal{L}$ to denote the tensor's gradient with respect to the loss function. Equation (6) is exactly the backpropagation approximation of Eq. (2), which is more conducive to optimization.

$$\frac{\partial \hat{X}_q}{\partial X_d} = X_d \tag{5}$$
$$\nabla_{X_d}\mathcal{L} = \nabla_{X_q}\mathcal{L} \odot X_d \tag{6}$$

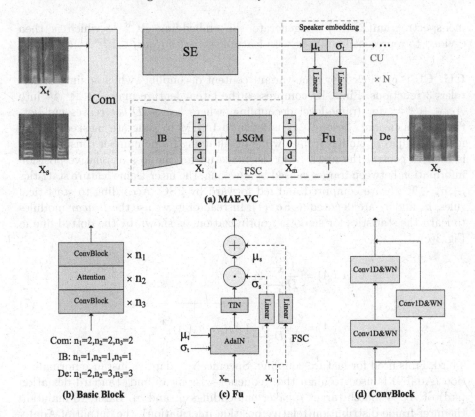

**Fig. 3.** The architecture of MAE-VC. Com, IB, and De follow the basic architecture of Basic Block except that the channel (frequency) dimension and the number of components are different. The processing of $\mu_t$ and $\sigma_t$ by *Linear* are simplified in (c). *WN* in (d) means weight normalization.

## 4.2  MAE-VC

In this section, we propose MAE-VC applying LSGM to transformation practice. The architecture shown in Fig. 3 consists of a shared encoder (Com), a speaker encoder (SE), $N$ cascaded units (CUs), and a decoder (De), where CU is comprised of an information bottleneck (IB), an LSGM and a feature fusion (Fu) module. There are 3 main improvements compared to other VC methods. First, LSGM is used for robust feature decoupling. Second, CU is used to enhance speaker style injection instead of the common explicit content encoder. Third, inter-frame feature statistics are fed forward by the feature-statistic connection (FSC, dashed line in Fig. 3) to supplement inter-frame relative position information.

**SE.** SE is derived from ADAINVC [4] and consists of cascaded *Conv*1ds for frame-level decoupling and *AvgPool*1ds for inter-frame compression. It accepts mel-spectrogram($\mathbb{R}^{80 \times L}$) to generate the embedding $\mathbb{R}^{256 \times 1}$, which is then divided to represent frequency-wise feature statistics ($\mu_t, \sigma_t \in \mathbb{R}^{128 \times 1}$).

**CU.** CU is employed for time-variant content decoupling, which mainly undertakes 3 functions. First, IB compresses the latent feature map $A \in \mathbb{R}^{D \times L}$ into $A_h \in \mathbb{R}^{bot \times L}$ for frame-level decoupling, where $bot \leq D$($bot$ represents bottleneck dimension). Second, $A_h$ is fed into LSGM to generate masked $X_m$ by masking adjacent similar frames while maintaining frame number consistency. $\theta$ is used to control the mask ratio. To make up for the loss of relative position information between frames caused by masking, the inter-frame feature statistics $\mu_s, \sigma_s \in \mathbb{R}^{1 \times L}$ are computed and fed forward by FSC. According to statistical rules, $\mu_s$ and $\sigma_s$ are denoted as Eq. (7). Instead of it, we use the *Linear* modules to learn the statistics for smoother optimization, as shown by the dotted line in Fig. 3(c).

$$\mu_s(A) = \frac{1}{D} \sum_{d=1}^{D} A_{dL}$$

$$\sigma_s(A) = \sqrt{\frac{1}{D} \sum_{d=1}^{D} (A_{dL} - \mu_s(A))^2 + \varepsilon} \tag{7}$$

Third, Fu is used for feature coupling. Specifically, adaptive instance normalization (AdaIN) is used to align the frequency-wise mean and standard deviation (std) of $A_h$ to match target speaker embeddings $\mu_t$ and $\sigma_t$. Later, to maintain the inter-frame distribution(relative position information), the output of AdaIN is normalized by TIN, and then multiplied by $\sigma_s$ and shifted by $\mu_s$. TIN represents transpose instance normalization. $y$ is the generic parameter. The main process can be denoted as Eq. (8).

$$TIN(y) = IN(y^T)^T$$

$$IN(A_h) = \frac{A_h - \mu_s(A_h)}{\sigma_s(A_h)}$$

$$AdaIN(A_h, \mu_t, \sigma_t) = IN(A_h) * \sigma_t + \mu_t$$

$$Fu(A_h, \mu_t, \sigma_t, \mu_s, \sigma_s) = TIN(AdaIN(A_h, \mu_t, \sigma_t)) * \sigma_s^T + \mu_s^T \tag{8}$$

**Training and Inferring.** Due to the lack of parallel data, MAE-VC performs the self-supervised reconstruction task for training, where source speech is the same as target speech($X_s = X_t$), shown in Eq. (9). $\mathcal{F}$ represents the VC function. Only $L1$ loss is required. In inferring phase, $X_s \neq X_t$. Converted speech $\hat{X}_c$ is generated as Eq. (11).

$$\hat{X}_s = \mathcal{F}(X_s, X_s) \tag{9}$$

$$\mathcal{L}_{rec} = \mathbb{E} \left\| X_s - \hat{X}_s \right\|_1 \tag{10}$$

$$\hat{X}_c = \mathcal{F}(X_s, X_t) \tag{11}$$

# 5 Experiments and Results

## 5.1 Experiment Conditions

**Preprocess.** For each utterance, the silent segments at the beginning and end are trimmed. The waveform samples are uniformly scaled to 0~0.95 to normalize volume. We extract 80-bin mel-spectrograms via STFT with FFT, window size, and hop size set to 1024, 1024, and 256. The mel-spectrograms are normalized to 0~1. The segments with length $L = 128$ are randomly selected for training.

**Implementation Details.** The training is performed on the VCTK corpus [28] with 58 speakers. The batch size is set to 100. We use AdamW for optimization, where $\beta 1 = 0.8$, $\beta 2 = 0.99$, and weight decay $\lambda = 0.00015$. CosineAnnealingLR with initial learning rate ($lr = 0.0001$) and learning rate decay factor ($lr\_d = 0.995$) is used to update the learning rate. Each model is trained for 400k steps. For inference, models are evaluated on four datasets: VCTK, VCC2020 [31], LibriSpeech [20], and Aishell-3 [24], each providing 1000 source-target pairs, except 280 for VCC2020. Aishell-3 is a Chinese dataset for cross-lingual evaluation. We use Base_Nc_botb_$\theta$r to represent the MAE-VC series where $N$ represents cascaded number, *bot* stands for bottleneck dimension and $\theta$ represents mask ratio. The hidden dimension in CU is uniformly set to 36. Further details may be found in our implementation code:[1]

## 5.2 Metrics

**Object Metrics.** 1) Speaker similarity accuracy(SAC). To evaluate the style similarity of the converted speech, we introduce the speaker verification system Dvector[2] to measure the speaker consistency between target speeches and converted speeches. SAC can be described as Eq. (12), where $N_p$ stands for the total number of sample pairs and $A$ represents the number of measured samples belonging to the same timbre.

$$SAC = \frac{A}{N_p} \qquad (12)$$

2) Word error rate (WER). To measure content consistency, two automatic speech recognition systems (wav2vec2[3] [1] for English; ASRT[4] for Chinese) are used to calculate WER based on the transcribed texts from source speeches and converted speeches. WER can be described as Eq. (13). We use $T_s$ and $T_c$ to denote the transcribe text from source speech and converted speech respectively. $N_s$ represents the total number of words in $T_s$. When converting $T_c$ to

---

[1] https://github.com/BrightGu/MAE-VC.
[2] https://github.com/yistLin/dvector.
[3] https://huggingface.co/docs/transformers/model_doc/wav2vec2.
[4] https://github.com/nl8590687/ASRT_SpeechRecognition.

$T_s$, $S$, $D$ and $I$ represent the number of substitutions, deletions and insertions respectively.

$$WER = \frac{S + D + I}{N_s} \tag{13}$$

**Subject Metrics.** The mean opinion score(MOS) test is performed on naturalness and similarity. For each dataset, the converted speeches are divided into 4 groups: F2F, F2M, M2F, and M2M(F for female and M for male, e.g., F2M represents female voice is converted to male voice). The samples are evaluated by 12 raters who are asked to assign a score of 1 5. The higher the score, the better the speaker similarity and naturalness. Statistical results are reported with 95% confidence intervals(CI). Our audio samples are available on the demo page[5].

## 5.3  Results and Discussions

**Evaluation on Mask Ratio and FSC.** The effects of mask ratio $\theta$ and FSC are investigated. The experiments are conducted on Base_4c_16b_$\theta$r with $\theta = 0, 30, 50, 70, 90$; applying FSC or not. As shown in Fig. 4, MAE-VC is robust to multiple English datasets and maintains high style similarity even on the Chinese dataset (Aishell-3). It can be inferred that adaptation to multiple datasets stems from self-supervised representation learning and robust feature decoupling in MAE-VC.

In terms of $\theta$, the impacts of $\theta$ on four datasets have maintained a consistent trend. Compared with $\theta = 0$, $\theta = 30$ can significantly improve the performance on style consistency with high SAC, which suggests that proper masking does help to remove source style information and extract robust content information. It can be further considered that LSGM effectively compresses redundant information between frames. However, as $\theta$ increases, the SAC exhibits a steady state, while the WER of the model without FSC consistently increases. In particular, the WER on Aishell-3 (Fig. 4(d)) even exceeds 1, which happens when the model performs poorly that the number of transcribed words exceeds the number of source words. The continued increase in WER illustrates that excessive compression caused by growing $\theta$ corrupts content information.

In fact, similarity-based masks can damage relative position information between frames, such as prosodic information. To compensate for the loss, the inter-frame statistical features are fed forward through FSC, which effectively reconstructs the inter-frame frequency energy distribution. Especially when $\theta = 90$, FSC reduces WER by about 50% on Aishell-3 without even affecting SAC. In conclusion, the combination of LSGM and FSC is pretty beneficial to the content-style trade-offs.

**Evaluation on Bottleneck.** The influence of *bot* in IB under the masking condition is investigated. Experiments are conducted on Base_4c_*bot*b_50r, where

---

[5] https://brightgu.github.io/MAE-VC/.

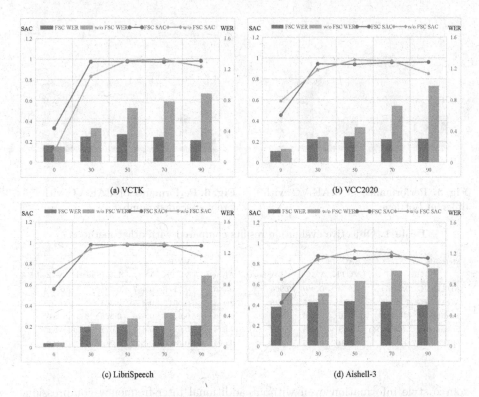

**Fig. 4.** Evaluations on four datasets using different mask rates, employing FSC or not. The histogram represents WER (right), and the line chart represents SAC (left).

$bot = 2, 4, 8, 16, 36$. The results in Fig. 5 indicate that $bot = 8$ almost achieves the highest SAC, although WER lags behind. In contrast, the widest $bot = 36$ achieves the best WER. However, the fluctuation of the metrics is almost less than 5% for each dataset, even though $bot$ is reduced by 18 times from 36 to 2. Compared with frame-level methods more sensitive to $bot$ (the performance of AGAINVC [2] can be significantly affected when $bot$ changed from 4 to 3), MAE-VC is highly adaptable to $bot$. The adaptation may originate from 2 aspects. First, since speech is a short-term stationary signal, the frequency energy distribution in a single frame has a unique periodic pattern that resonant frequencies are mostly derived from the fundamental frequency. Regardless of inter-frame compression, for each frame, the close inter-frequency correlation means redundant high-dimensional information can be compressed into a low-dimensional representation through a narrow bottleneck without information loss. Therefore, in the case of FSC, $bot = 2$ is highly likely to be acceptable. Second, we can infer that for content decoupling, the effect of inter-frame compression is more significant than that of inter-frequency compression. As the time-invariant style information in the source speech is important redundant information, the LSGM-based inter-frame compression method can more effectively remove the

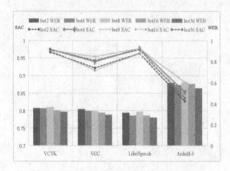

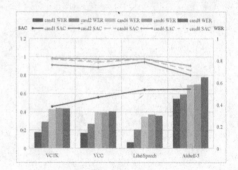

**Fig. 5.** Performance of MAE-VC with different bottlenecks.

**Fig. 6.** Performance of MAE-VC with different cascaded number.

**Table 1.** Objective evaluation results compared with other methods.

| Methods | SAC | | | | WER | | | |
|---|---|---|---|---|---|---|---|---|
| | VCTK | VCC | LibriSpeech | Aishell-3 | VCTK | VCC | LibriSpeech | Aishell-3 |
| ADAINVC (9.04M) | 0.972 | **0.996** | 0.966 | **0.925** | 0.717 | 0.497 | 0.568 | 0.874 |
| AGAINVC (7.93M) | 0.878 | 0.865 | 0.828 | 0.478 | 0.551 | 0.376 | 0.344 | 0.769 |
| MediumVC (26.41M) | 0.976 | 0.946 | 0.876 | 0.698 | 0.535 | 0.626 | 0.578 | 0.816 |
| FragmentVC (48.01M) | 0.958 | 0.935 | 0.902 | 0.874 | 0.624 | 0.641 | 0.671 | 0.845 |
| MAE-VC (7.94M) | **0.985** | 0.975 | **0.977** | 0.904 | **0.365** | **0.327** | **0.308** | **0.584** |

source style information even without additional inter-frequency compression. For example, ADAINVC [4] achieves competitive style consistency with 16x inter-frame compression and only 4x inter-frequency compression.

**Evaluation on Cascaded Number.** We use Base_$N$c_16b_50r to conduct the evaluation, where $N = 1, 2, 4, 6, 8$, and the parameters are 6.68–8.37M. Results in Fig. 6 show that when $N \leq 4$, SAC grows with increasing $N$, meanwhile WER also increases gradually. It suggests that increasing CU is beneficial to the injection of target style information, though damages some content information. When $N \geq 4$, the performances tend to be stable, which suggests that MAE-VC based on LSGM does not suffer from over-compression caused by excessive cascaded processing. It can be inferred that when $N$ is small, the similarity-based masking strategy can effectively remove redundant information, and when $N$ is large, it can also dynamically protect the core content information from being destroyed. Furthermore, we note that the performance of the models is not significantly correlated with the number of parameters in the case of similar structures, which also motivates our future exploration of lightweight models.

**Compared with Other Methods.** We adopt base_6c_16b_50r to compare with the present four models: ADAINVC [4], AGAINVC [2], MediumVC [6] and FragmentVC [18]. Table 1 and Table 2 show the objective and subjective results on four datasets, respectively. ADAINVC is a VAE-based model, which

**Table 2.** Subjective evaluation results compared with other methods. MOS stands for the metrics on naturalness and SMOS for style similarity. $\pm$ represents the fluctuation of the value with 95% confidence intervals.

| Methods | SMOS | | | | MOS | | | |
|---|---|---|---|---|---|---|---|---|
| | VCTK | VCC | LibriSpeech | Aishell-3 | VCTK | VCC | LibriSpeech | Aishell-3 |
| ADAINVC | $3.57\pm0.03$ | $3.58\pm0.03$ | $3.52\pm0.02$ | $3.33\pm0.02$ | $3.47\pm0.03$ | $3.46\pm0.02$ | $3.47\pm0.03$ | $2.97\pm0.03$ |
| AGAINVC | $3.40\pm0.02$ | $3.52\pm0.03$ | $3.38\pm0.03$ | $3.11\pm0.03$ | $3.52\pm0.02$ | $3.54\pm0.03$ | $3.51\pm0.04$ | $3.15\pm0.03$ |
| MediumVC | $3.54\pm0.02$ | $3.57\pm0.02$ | $3.47\pm0.03$ | $3.27\pm0.03$ | $3.51\pm0.03$ | $3.53\pm0.03$ | $3.41\pm0.03$ | $3.12\pm0.02$ |
| FragmentVC | $3.46\pm0.03$ | $3.54\pm0.03$ | $3.45\pm0.03$ | $3.24\pm0.03$ | $3.46\pm0.04$ | $3.42\pm0.03$ | $3.40\pm0.03$ | $2.85\pm0.03$ |
| MAE-VC | $\mathbf{3.71\pm0.03}$ | $\mathbf{3.61\pm0.03}$ | $\mathbf{3.56\pm0.03}$ | $\mathbf{3.42\pm0.04}$ | $\mathbf{3.64\pm0.04}$ | $\mathbf{3.62\pm0.02}$ | $\mathbf{3.54\pm0.03}$ | $\mathbf{3.30\pm0.03}$ |

adopts a ratio-based method for content decoupling, where the frame dimension is compressed 16 times and the frequency dimension is compressed 4 times. The high SAC and SMOS indicate ADAINVC based on ratio-based decoupling achieves excellent style expression. However, competitive style consistency may be achieved by sacrificing content consistency. Likewise, MAE-VC uses a similar compression ratio, masking half of the frames each time in four successive compressions. In contrast, MAE-VC achieves close style consistency and significantly improves content consistency. AGAINVC is an Unet-based model, which employs the frame-level decoupling ($bot = 4$). Compared with MAE-VC and ADAINVC, results show that although frame-level decoupling is advanced in preserving content, it is not robust or flexible in removing source style information. It can be inferred for content decoupling, the frame-level method is indeed not as efficient as the utterance-level method. MediumVC adopts pre-trained Dvector to represent speaker embeddings, while FragmentVC uses pre-trained Wav2Vec2 to represent content structures. However, both fail to achieve competitive content-style trade-offs. In conclusion, MAE-VC uses LSGM to process the inter-frame redundant information to obtain a competitive style expression and introduces FSC to compensate for the disordered inter-frame statistical features caused by the mask, achieving significant content-style trade-offs.

# 6    Conclusion

In this work, we deeply explore and analyze the strengths and weaknesses of feature decoupling methods in common VC methods. On this basis, we propose LSGM, which is considered a self-referential method that measures the similarity between adjacent frames through neural networks and removes redundant information between frames by masking similar frames. MAE-VC is designed based on LSGM, where FSC is further introduced to compensate for the disordered inter-frame relative position information. The superiority of LSGM prompts us to rethink the kernel of feature decoupling that robust performance should come from the in-depth exploration of intrinsic correlations of speech features, rather than unexplained deep features. Furthermore, we find that robust performance is not parameter-sensitive, but more structure-dependent. In future work, we will continue to explore lightweight methods to facilitate the application of VC.

# References

1. Baevski, A., Zhou, Y., Mohamed, A., Auli, M.: Wav2Vec 2.0: a framework for self-supervised learning of speech representations. Adv. Neural Inf. Process. Syst. **33**, 12449–12460 (2020)
2. Chen, Y.H., Wu, D.Y., Wu, T.H., Lee, H.Y.: Again-VC: a one-shot voice conversion using activation guidance and adaptive instance normalization. In: ICASSP 2021–2021 IEEE International Conference on Acoustics, Speech and Signal Processing (ICASSP), pp. 5954–5958. IEEE (2021)
3. Choi, Y., Choi, M., Kim, M., Ha, J.W., Kim, S., Choo, J.: StarGAN: unified generative adversarial networks for multi-domain image-to-image translation. In: Proceedings of the IEEE Conference on Computer Vision and Pattern Recognition, pp. 8789–8797 (2018)
4. Chou, J.C., Yeh, C.C., Lee, H.Y.: One-shot voice conversion by separating speaker and content representations with instance normalization. arXiv preprint arXiv:1904.05742 (2019)
5. Erro, D., Moreno, A., Bonafonte, A.: Voice conversion based on weighted frequency warping. IEEE Trans. Audio Speech Lang. Process. **18**(5), 922–931 (2009)
6. Gu, Y., Zhang, Z., Yi, X., Zhao, X.: MediumVC: any-to-any voice conversion using synthetic specific-speaker speeches as intermedium features. arXiv preprint arXiv:2110.02500 (2021)
7. He, K., Chen, X., Xie, S., Li, Y., Dollár, P., Girshick, R.: Masked autoencoders are scalable vision learners. In: Proceedings of the IEEE/CVF Conference on Computer Vision and Pattern Recognition, pp. 16000–16009 (2022)
8. Helander, E., Silén, H., Virtanen, T., Gabbouj, M.: Voice conversion using dynamic kernel partial least squares regression. IEEE Trans. Audio Speech Lang. Process. **20**(3), 806–817 (2011)
9. Hsu, C.C., Hwang, H.T., Wu, Y.C., Tsao, Y., Wang, H.M.: Voice conversion from non-parallel corpora using variational auto-encoder. In: 2016 Asia-Pacific Signal and Information Processing Association Annual Summit and Conference (APSIPA), pp. 1–6. IEEE (2016)
10. Hsu, C.C., Hwang, H.T., Wu, Y.C., Tsao, Y., Wang, H.M.: Voice conversion from unaligned corpora using variational autoencoding wasserstein generative adversarial networks. arXiv preprint arXiv:1704.00849 (2017)
11. Huang, X., Belongie, S.: Arbitrary style transfer in real-time with adaptive instance normalization. In: Proceedings of the IEEE International Conference on Computer Vision, pp. 1501–1510 (2017)
12. Jia, Y., et al.: Transfer learning from speaker verification to multispeaker text-to-speech synthesis. Adv. Neural Inf. Process. Syst. **31**, 4485–4495 (2018)
13. Kameoka, H., Kaneko, T., Tanaka, K., Hojo, N.: ACVAE-VC: non-parallel many-to-many voice conversion with auxiliary classifier variational autoencoder. arXiv preprint arXiv:1808.05092 (2018)
14. Kameoka, H., Kaneko, T., Tanaka, K., Hojo, N.: StarGAN-VC: non-parallel many-to-many voice conversion using star generative adversarial networks. In: 2018 IEEE Spoken Language Technology Workshop (SLT), pp. 266–273. IEEE (2018)
15. Kaneko, T., Kameoka, H.: CycleGAN-VC: non-parallel voice conversion using cycle-consistent adversarial networks. In: 2018 26th European Signal Processing Conference (EUSIPCO), pp. 2100–2104. IEEE (2018)
16. Kingma, D.P., Welling, M.: Auto-encoding variational bayes. arXiv preprint arXiv:1312.6114 (2013)

17. Lee, S.H., Kim, J.H., Chung, H., Lee, S.W.: VoiceMixer: adversarial voice style mixup. Adv. Neural. Inf. Process. Syst. **34**, 294–308 (2021)
18. Lin, Y.Y., Chien, C.M., Lin, J.H., Lee, H.Y., Lee, L.S.: FragmentVC: any-to-any voice conversion by end-to-end extracting and fusing fine-grained voice fragments with attention. In: ICASSP 2021–2021 IEEE International Conference on Acoustics, Speech and Signal Processing (ICASSP), pp. 5939–5943. IEEE (2021)
19. Luan, Y., Saito, D., Kashiwagi, Y., Minematsu, N., Hirose, K.: Semi-supervised noise dictionary adaptation for exemplar-based noise robust speech recognition. In: 2014 IEEE International Conference on Acoustics, Speech and Signal Processing (ICASSP), pp. 1745–1748. IEEE (2014)
20. Panayotov, V., Chen, G., Povey, D., Khudanpur, S.: LibriSpeech: an ASR corpus based on public domain audio books. In: 2015 IEEE International Conference on Acoustics, Speech and Signal Processing (ICASSP), pp. 5206–5210. IEEE (2015)
21. Qian, K., Zhang, Y., Chang, S., Yang, X., Hasegawa-Johnson, M.: AutoVC: zero-shot voice style transfer with only autoencoder loss. In: International Conference on Machine Learning, pp. 5210–5219. PMLR (2019)
22. Ren, Y., et al.: FastSpeech 2: fast and high-quality end-to-end text to speech. arXiv preprint arXiv:2006.04558 (2020)
23. Ronneberger, O., Fischer, P., Brox, T.: U-net: convolutional networks for biomedical image segmentation. In: Navab, N., Hornegger, J., Wells, W.M., Frangi, A.F. (eds.) MICCAI 2015. LNCS, vol. 9351, pp. 234–241. Springer, Cham (2015). https://doi.org/10.1007/978-3-319-24574-4_28
24. Shi, Y., Bu, H., Xu, X., Zhang, S., Li, M.: Aishell-3: a multi-speaker mandarin tts corpus and the baselines. arXiv preprint arXiv:2010.11567 (2020)
25. Tian, X., Lee, S.W., Wu, Z., Chng, E.S., Li, H.: An exemplar-based approach to frequency warping for voice conversion. IEEE/ACM Trans. Audio Speech Lang. Process. **25**(10), 1863–1876 (2017)
26. Toda, T., Black, A.W., Tokuda, K.: Voice conversion based on maximum-likelihood estimation of spectral parameter trajectory. IEEE Trans. Audio Speech Lang. Process. **15**(8), 2222–2235 (2007)
27. Ulyanov, D., Vedaldi, A., Lempitsky, V.: Instance normalization: the missing ingredient for fast stylization. arXiv preprint arXiv:1607.08022 (2016)
28. Veaux, C., Yamagishi, J., MacDonald, K., et al.: Superseded-CSTR VCTK corpus: English multi-speaker corpus for CSTR voice cloning toolkit (2016)
29. Wan, L., Wang, Q., Papir, A., Moreno, I.L.: Generalized end-to-end loss for speaker verification. In: 2018 IEEE International Conference on Acoustics, Speech and Signal Processing (ICASSP), pp. 4879–4883. IEEE (2018)
30. Wu, D.Y., Chen, Y.H., Lee, H.Y.: VQVC+: one-shot voice conversion by vector quantization and U-net architecture. arXiv preprint arXiv:2006.04154 (2020)
31. Zhao, Y., et al.: Voice conversion challenge 2020: intra-lingual semi-parallel and cross-lingual voice conversion. arXiv preprint arXiv:2008.12527 (2020)
32. Zhu, J.Y., Park, T., Isola, P., Efros, A.A.: Unpaired image-to-image translation using cycle-consistent adversarial networks. In: Proceedings of the IEEE International Conference on Computer Vision, pp. 2223–2232 (2017)

# Visual Explanations for Exposing Potential Inconsistency of Deepfakes

Pengfei Pei[1,2], Xianfeng Zhao[1,2(✉)], Yun Cao[1,2], and Chengqiao Hu[1]

[1] State Key Laboratory of Information Security, Institute of Information
Engineering, Chinese Academy of Sciences, Beijing 100195, China
{peipengfei,zhaoxianfeng,caoyun,huchengqiao}@iie.ac.cn
[2] School of Cyber Security, University of Chinese Academy of Sciences,
Beijing 100195, China

**Abstract.** In recent years, the rapid development of Deepfake has aroused public concerns. Existing Deepfake detection methods mainly focus on improving the accuracy. However, when real-world victims require additional interpretable results to refute, the accuracy of these methods is certainly insufficient. To mitigate this issue, we delve into forgery traces and propose a novel framework, named Find-X, that presents additional visual information as an explanation of the results. Specifically, we design a new module named Separation Potential Inconsistency (SPI) which aims to visually explain the forgery traces of fake videos. Find-x detection of Deepfake consists of three stages: (1) A frequency-aware module and a spatial-aware module to enhance the features. (2) A multi-scale feature extraction module to extract richer features. (3) A classification module and a SPI module to output the visual explanations. Our method outperforms state-of-the-art competitors on three popular benchmark datasets: FaceForensics++, Celeb-DF, and DeepFakeDetection. In addition, extensive visualization experiments on FaceForensics++ demonstrate that SPI can effectively separate the potentially inconsistent features of videos generated by five different Deepfake methods.

**Keywords:** Deepfake detection · Visual explanations · Interpretability study

## 1 Introduction

Since Deepfake has become a popular forgery generation tool, a large number of malicious fake videos have caused great loss of personal reputation and property. As the quality of generated Deepfake videos improves, the difficulty of Deepfake detection increases [9,14,31,38]. The focus of the research has been changed from the improvement of accuracy to the improvement of detection robustness [3,17,34,35,39].

Currently, most of the research focuses on improving the generality on unknown datasets. However, although these methods improve the robustness

X. Zhao et al. (Eds.): IWDW 2022, LNCS 13825, pp. 68–82, 2023.
https://doi.org/10.1007/978-3-031-25115-3_5

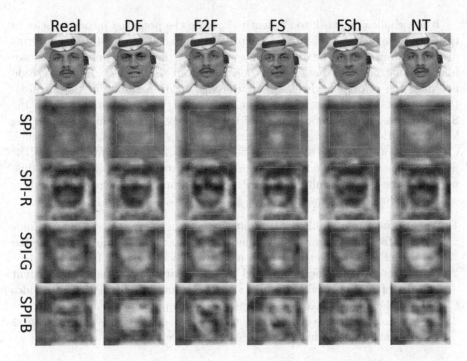

**Fig. 1.** Visual explanations for exposing potential inconsistencies in Deepfake, evaluated on FaceForensics++ of multi classification. The marked region in the SPI figure corresponds to the region of face forgery. SPI-R, SPI-G and SPI-B explain the potential inconsistencies from different perspectives (edges, pixels, areas). The edges, pixels and regions of the SPI of the real face are smooth and the facial features are sharp. Conversely, the edges of the fake face do not coincide with the edges of the original face, the pixel continuity is inconsistent, and the distribution of statistical features in the region is not uniform. The facial features of SPI (R, G, B) have obvious defects, especially the mouth and nose parts.

of the detection, they are not sufficient to provide a powerful countermeasure for the victim. Victims generally need clues from fake videos to give evidence. Synthetic videos typically have significantly inconsistent information between the statistics of the forgery region and the original region. Based on this observation, we propose a new vision transformer (ViT) based framework named Find-X. Find-X is designed as a multi-stage structure that extracts both coarse-grained features and fine-grained features to fully detect such forgery traces. Find-X contains an innovatively designed module named separation potentially inconsistency (SPI). As shown in Fig. 1, the SPI generates an additional visual explanations including edges, pixels, and areas that explain the inconsistency between the forgery and the original regions. SPI provides the victims with a alternative way to proof their innocence. Meanwhile, the inconsistency features can effectively improve the robustness of detection [11,39].

It is a challenging task to efficiently separate the potential inconsistencies in the forgery region of Deepfake and express the learned inconsistencies in terms of visual information. The difficulty lies in the fact that we do not have a known function that can feed back SPI by learning from the dataset. Because we do not have the priori knowledge about the visualization results of the inconsistent features. Our idea is to share the feature extraction parameters with the SPI and classification modules. We strive to make SPI faithfully express the learned multi-scale features and accurately correlate the classification results. Continuous optimization of the classification module can potentially affect the results of the SPI module. In other words, the better the classification module performs, the better results of the SPI module would have. Our contributions can be concluded as follows:

1. We propose a novel multi-scale, multi-stage framework named Find-X. Find-X uses a frequency-aware module and a spatial-aware module to enhance the features of fake videos that can be used to efficiently detect Deepfake.
2. We propose a novel module named SPI to separate potentially inconsistent features of Deepfake videos. The visual explanations of SPI can provides additional evidence to expose the essence of Deepfake better.
3. We carry out extensive experiments to verify the effectiveness and robustness of our method, and achieve state-of-the-art detection performance.

## 2   Related Works

### 2.1   Deepfake Detection

In recent years, numerous Deepfake detection methods have been proposed. Some detection methods focus on optimizing network structure to improve Deepfake detection accuracy [13,17,28,38]. The others use the inconsistencies of forgery faces to detect videos [7,11,18,39]. The researchers of the former approaches used some simple CNN structures. With the development of deep learning technology, advanced network structures such as Xception, Long Short-Term Memory (LSTM) and ViT have been applied to Deepfake detection and achieved better performance. The methods are data-driven [3,20,35], and do not rely on any specific artifact. Some later methods use faked video inconsistencies in visual, audio or frames continuity to detect fake videos, such as the LipForensics [11] make use of high-level semantic irregularities in mouth movements, which are common in numerous generated videos. Moreover, those videos with such low visual qualities can be definitely distinguished by the human eyes. At present, the focus of research has been shifted from improving the accuracy to providing interpretable evidenced [18,24,31].

### 2.2   Vision Transformer

At present, the ViT-based networks have shown great success in a wide range of domains including image and video tasks [1,4,15,21,36]. ViT is an effective

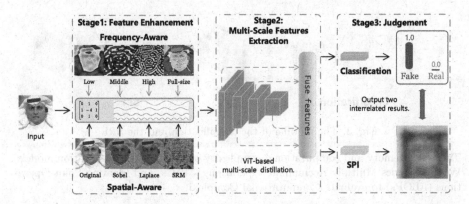

**Fig. 2.** The framework of Find-X has three stages that play an essential role in our framework. **Stage1:** the frequency-aware module and the spatial-aware module to enhance the features of the fake videos. **Stage2:** the ViT-based multi-scale feature extractor to learn richer characteristics. **Stage3:** the classification module and the SPI module that can explain the result visually.

feature extraction structure for video, especially for sequence-to-sequence modeling [30,32,33]. ViViT [1] presents pure-transformer based models for video classification that use tokens to extract spatio-temporal features of videos more efficiently. ViT provides an alternative besides 3D CNN and mixed 2D CNN for video understanding tasks. In early studies, the ViT-based models are known to be effective only when large training datasets are available. The parameters and calculation cost of ViT structure increase exponentially with the increase of image pixels (same patch size). To solve this problem, researchers proposed various improved ViT architectures. Swin Transformer [22] shifted windowing scheme brings greater efficiency by limiting self-attention computation to non-overlapping local windows while also allowing cross-window connection. This hierarchical architecture has the flexibility to model at various scales and has linear computational complexity with respect to image size. HRFormer [36] take advantage of the multi-resolution parallel design introduced in high-resolution convolutional networks, along with local-window self-attention that performs self-attention over tiny non-overlapping image windows, for improving the memory and computation efficiency. Recent studies have combined CNN and ViT and achieved better performance [4,10].

## 3   Method

Find-X is a frame-level detection method. As shown in Fig. 2, Find-X consists of three stages: (1) Feature enhancement, we design a frequency-aware module and a spatial-aware module to enhance the inconsistent features of videos, which is conducive to improving the result of detection. (2) Multi-Scale feature extraction, we design a multi-scale feature extractor to learn richer features. (3) Judgement,

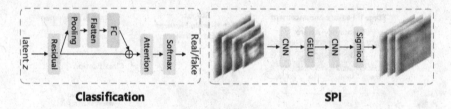

**Classification**                                          **SPI**

Fig. 3. The module of the classification and the SPI.

Table 1. Memory and computational complexity of widely used classification models. We use Metrics Multiply Accumulate Operations (MACs) and Floating-Point Operations (FLOPs) to evaluate Computational Complexity.

| Model | Image size | FLOPs (G) | MACs (G) | Params (M) |
|---|---|---|---|---|
| PoolingVisionTransformer | 224 × 224 | 21.1 | 10.55 | 73.76 |
| HRNet (w44) | 224 × 224 | 29.94 | 14.97 | 67.06 |
| Twins-PVT (base) | 224 × 224 | 12.9 | 6.45 | 43.83 |
| SwinTransformer | 224 × 224 | 30.28 | 15.14 | 87.77 |
| Xcepetion | 224 × 224 | 9.18 | 4.59 | 20.81 |
| resnet50 | 224 × 224 | 8.24 | 4.12 | 25.56 |
| Find-X (Ours) | 224 × 224 | 43.78 | 21.89 | 119.45 |

we design a classification module to output the result and a SPI module to explain the result. As shown in Table 1, Find-X is a larger model than the commonly used classification model to improve the accuracy of detection. The details of our approach will be demonstrated below.

### 3.1 Feature Enhancement

**Face Preprocessing.** The main goal of face preprocessing is to extract faces from each frame in the video and then improve the accuracy of Deepfake detection. Face preprocessing consists of face detection and face selection. As a result, the detection accuracy can be effectively improved by performing the proposed preprocessing method.

**Face Detection.** We use the open-source face detection tool MTCNN [37]. MTCNN is a widely used face detection method with high accuracy, fast running speed and smaller memory occupation. We use MTCNN to extract the face of each frame in the video. Before training, we extract faces from the video in the dataset and save them as pictures in lossless format. During the testing, each video also requires face extraction in advance.

**Face Selection.** Given a fake video, the fake faces to be detected is defined as $\mathbb{F} = \gamma \times t$, where $\gamma$ is the frame rate and $t$ is the video duration (seconds). Detecting whether the video is fake is equivalent to calculating the average forgery probability $x \in \mathbb{F}$. The cost of calculating all $\mathbb{F}$ is huge, and the low quality face also reduces the accuracy of detection. Therefore, it is necessary to select the helpful $x$. We select faces from two perspectives: face quality $q$ and face pixel size $s$. The face quality $q$ (0–1) is the probability value of MTCNN while judging whether the image contains a face. In general, large $q$ indicates superior facial quality. The $s$ is the pixel width of the face in the image. According to the traditional experimental experience of face detection and the face resolution in the video of the datasets, we set face pixel size $s > 80$ and face quality $q > 0.85$ in the experiments.

**Spatial-Aware Enhancement.** We observe that the fake video has distinct traces on the edges of the forgery object. Therefore, we use Sobel operator [16] to detect edges of fake videos more effectively. We then additionally use the Laplacian operator to detect edges, since the Laplacian operator [27] is robust to rotation. We represent the x and y operators as two CNN $3 \times 3$ convolution kernels that extract features from horizontal and vertical directions of videos.

We also observe that the forgery region has different statistical characteristics from the original region. Based on these observations, we find that some artificially designed spatial operators can considerably enhance these forgery traces. We use the Spatial Rich Model (SRM) [6] operator to enhance region detection of fake videos because the SRM operator is sensitive to continuous statistical features of pixels. We use three $5 \times 5$ operators to respectively filter the R, G, B channels of the videos, which can better characterize the damage of forgery to the multi-correlation of neighborhood pixels.

All operators are implemented by CNN based filters, which replace the kernels of CNN filters with these operators. We also normalize the operators to 0–1 to help better extract features.

**Frequency-Aware Enhancement.** We use PyTorch to implement DCT based CNN filters to extract frequency features of videos. We use the DCT filters of three different sizes: high, middle and low to learn the local DCT features of the video, and a full-size DCT filter to learn the overall features of the video. Mostly, DCT is used in data or image compression to convert spatial signals into frequency domain. In particular, DCT has excellent correlation properties, which makes it easier to detect forgery traces that are undetectable in the spatial domain.

The multi-size frequency-aware module decomposes the image into different frequency domain features. Moreover, the corresponding features are learned with learnable CNNs. These frequency domain features are learned from different receptive fields of different sizes. The learned traces on the frequency domain features reveal the irrelevance of the forged region. This will provide an important basis for SPI and classification results.

## 3.2   Multi-scale Feature Extraction

We implement a multi-stage and multi-scale feature extraction network based on PoolFormer [4]. PoolFormer uses a pooling module instead of an attention module, which combines the advantages of Convolutional Neural Networks (CNN) with less parameters and the ViT sequence-friendly. The PoolFormer can effectively use various meta-information to improve the performance of fine-grained recognition. The two types of Spatial-Aware and Frequency-Aware features are gradually learned from coarse to fine in a pyramid structure. The multi-level feature extraction network guarantees the effectiveness of features.

## 3.3   Classification

As shown in Fig. 3, we designed a classification module that combines CNNs and attention mechanism. First, multi-scale features are fused with residual CNN blocks. Then an attention module is used to help learning of the features. Finally, the probability value is outputted through a softmax function.

## 3.4   SPI

**Component Design.** While the classification module can use the Binary Cross Entropy Loss as a feedback, the SPI module can only be evaluated subjectively. The classification module feeds back the learning results to the shared feature extraction parameters. A shallower network is conducive to the SPI module to associate with the feature extraction parameters. Therefore, we designed a shallower SPI module to fewer interfere with the results of visual inconsistent information. As shown in Fig. 3, the SPI consists of multiple layers of CNN blocks with different structures and GELU functions. Finally, the RGB image is outputted through a sigmoid function. The SPI module is used to represent the forgery edges, pixels and areas.

**Visual Explanations.** We learn the inconsistent feature through the spatial-aware module to find out the inconsistencies between forgery region and the original region in terms of edge and pixel continuity. At the same time, we learn the uncorrelation of the forgery region through the frequency-aware module. The inconsistent information in the spatial and frequency domain is fully learned through the multi-scale feature extraction module. As a result, SPI visually expresses potentially inconsistent information from edges, pixels and regions. The training feedback process of SPI is associated with the classification module, therefore better classification result leads to better visual expression.

## 4   Experiments

In this section, we evaluate our method in two aspects. In the first aspect, we compare the performance of our method with recent methods. In the second aspect, we evaluate the visualization performance of SPI on several widely used datasets.

**Table 2.** Comparison with related methods on FaceForensics++ dataset with High-Quality (HQ) and Low-Quality (LQ) by training on FaceForensics++ raw.

| Method | LQ | | HQ | |
|---|---|---|---|---|
| | ACC | AUC | ACC | AUC |
| Face X-ray [18] | – | 0.616 | – | 0.874 |
| Xception [2] | 0.868 | 0.893 | 0.957 | 0.963 |
| F3-Net [25] | 0.904 | 0.933 | 0.975 | 0.981 |
| EfficientNet-B4 [29] | 0.866 | 0.882 | 0.966 | 0.991 |
| MA(Xception) [38] | 0.869 | 0.872 | 0.963 | 0.989 |
| MA(Efficient-B4) [38] | 0.886 | 0.904 | 0.976 | 0.992 |
| Ours | **0.991** | **1.0** | **0.998** | **1.0** |

**Table 3.** Comparison with state-of-the-art methods on three public datasets: Celeb-DF, DeepFakeDetection and Deepfakes of FaceForensics++.

| Method | Celeb-DF | DeepFakeDetection | FaceForensics++ (Deepfakes) |
|---|---|---|---|
| Xception [2] | 0.994 | – | 0.955 |
| DILNet [9] | 0.996 | – | 0.981 |
| Grad-CAM [23] | 0.794 | 0.919 | 0.992 |
| DIANet [14] | – | – | 0.904 |
| STIL [8] | 0.996 | – | 0.971 |
| FInter [12] | 0.905 | – | 0.957 |
| ViTHash [24] | 0.994 | 0.963 | 0.999 |
| Ours | **0.999** | **1.0** | **1.0** |

## 4.1 Experiment Setting

**Datasets.** We evaluate Find-X on several widely used datasets: FaceForensics++, DeepFakeDetection and Celeb-DF. The details of datasets are described as follows:

1. **FaceForensics++.** The FaceForensics++ [26] dataset contains 1000 real videos from YouTube with over 500,000 frames. As well as the same number of manipulated videos generated by various of the state-of-the-art Deepfake methods. Moreover, the ratio of tampered videos and original videos is 1:1.
2. **DeepFakeDetection.** The Google/Jigsaw Deepfake detection (DeepFakeDetection) dataset has [5] over 363 original videos from 28 consented actors of various genders, ages and ethnic groups. The DeepFakeDetection dataset contains 3,068 fake videos generated by four basic Deepfake methods. The ratio of original videos and fake videos is 0.12:1.
3. **Celeb-DF.** The Full Facebook DeepFake detection challenge [19] dataset is part of the DeepFake detection challenge, composed of 590 real videos and 5,639 fake videos. The ratio of synthetic clips and real clips is 1:0.23.

**Table 4.** Evaluation on FaceForensics++ with five different forgery methods by training on FaceForensics++ raw for multi classification.

| Compression | Training/Test set (ACC) | | | | | |
|---|---|---|---|---|---|---|
| | Real | Deepfakes | Face2Face | FaceSwap | NeuralTextures | FaceShifter |
| Raw | 0.994 | 0.993 | 0.990 | 0.994 | 0.991 | 0.993 |
| C23 | 0.978 | 0.995 | 0.988 | 0.995 | 0.990 | 0.995 |
| C40 | 0.318 | 0.998 | 0.967 | 0.991 | 0.990 | 0.998 |

**Table 5.** Evaluation on FaceForensics++ with five different forgery methods by training on FaceForensics++ raw for binary classification.

| Compression | Training/Test set (ACC) | | | | |
|---|---|---|---|---|---|
| | Deepfakes | Face2Face | FaceSwap | NeuralTextures | FaceShifter |
| Raw | 1.0 | 0.998 | 1.0 | 1.0 | 1.0 |
| C23 | 1.0 | 0.998 | 1.0 | 0.994 | 1.0 |
| C40 | 1.0 | 0.999 | 1.0 | 0.957 | 1.0 |

**Implementation Details.** Our models are implemented by PyTorch, and the code has been released to GitHub. We use ffmpeg to segment the video into frames, and train models with a single NVIDIA RTX 3090 24 GB GPU card. Each model of the dataset is trained for 2–5 epochs. Additionally, we use the Adaptive Moment Estimation (ADAM) optimizer with a learning rate of $1e-5$. Being computationally efficient, ADAM requires less memory and outperforms on large datasets.

**Baseline Methods.** To verify the performance of our approach, we compare it with six recent works: the Face X-ray [18], Xception [2], F3-Net [25], EfficientNet-B4 [29], MA(Xception) [38] and MA(Efficient-B4) [38] on FaceForensics++. We also compare our method with six state-of-the-art methods: Xception [26], DIL-Net [9], Grad-CAM [23], DIANet [14], STIL [8], FInter [12] and ViTHash [24] on Celeb-DF, DeepFakeDetection and FaceForensics++ (Deepfakes).

## 4.2 Comparison Results

We first compare the performance of the videos with state-of-the-art methods on the FaceForensics++ dataset with different video quality. To avoid any unnecessary misunderstanding, the results of the comparative methods are directly cited from [38]. As shown in Table 2, experimental results show that our method is superior to related methods, especially in the LQ setting our method has obvious advantages. This is mainly due to the significant loss of textural information caused by high compression, but the proposed Find-X effectively reduces the effect of video compression by enhancing spatial and frequency features.

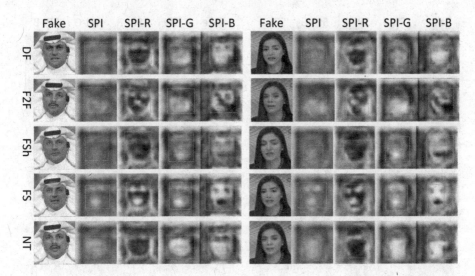

**Fig. 4.** Visual evaluation of our approach on FaceForensics++ with binary classification.

We then conduct comparison experiments with several recent works on the widely used datasets Celeb-DF, DeepFakeDetection and FaceForensics++. On DeepFakeDetection we train on c23 and test on c40. On FaceForensics++ we train on raw and test on c23. As shown in Table 3, we achieve state-of-the-art performance on Celeb-DF. We also achieve 100% accuracy on DeepFakeDetection and FaceForensics++, which is a significant advantage over comparable methods.

### 4.3 Distinguish Different Forgeries

We conduct extensive experiments on FaceForensics++ to distinguish different forgeries. It is more challenging to detect different forgery methods than the binary classification task of judging whether the forgery is real. As shown in Table 4, the average accuracy of our method for distinguishing different types of video is 95.3%, and the maximum performance is 99.8%. It demonstrates that our method can accurately distinguish the subtle differences between different types of forged videos.

### 4.4 Robustness to Video Compression

In this part, we primarily verify the robustness of Find-X with video compression. We evaluate the accuracy of three different compression ratios (raw, c23 and c40) on five subsets of FaceForensics++. As shown in Table 4 and Table 5, with the increase of compression rate, the accuracy of our method decreases a little. As shown in Table 2, our method significantly outperforms the existing methods on

| Fake | SPI | SPI-R | SPI-G | SPI-B | Fake | SPI | SPI-R | SPI-G | SPI-B |
|------|-----|-------|-------|-------|------|-----|-------|-------|-------|

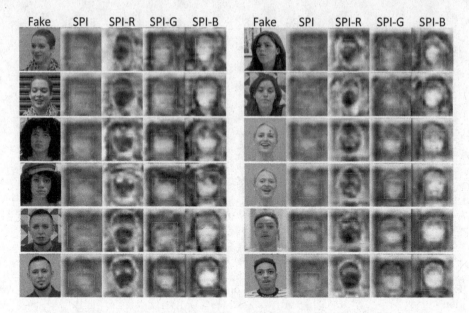

**Fig. 5.** Visual evaluation of our approach on DeepFakeDetection.

low compression videos. These experiments demonstrate that Find-X has good robustness to the video compression.

### 4.5 Visualization Analysis

In this part, we aim to separate the potentially inconsistent features of the forgery region by an additional designed module SPI, and interpret them with visual information. The results of the SPI module is shown in Fig. 5 and 6. Highlights of the image in the SPI that localizes the suspected area: (1) The SPI in the picture is a visual explanation of Deepfakes detection. (2) The SPI-R is most likely to detect the edge of the fake region. (3) The SPI-G is most likely to detect the continuous of pixels of the all picture. (4) The SPI-B is most likely to detect the region of the fake region. As shown in Fig. 4, the visual inconsistency of different forgery methods is significantly different. Experimental results show that our designed module SPI can effectively separate inconsistencies in fake videos.

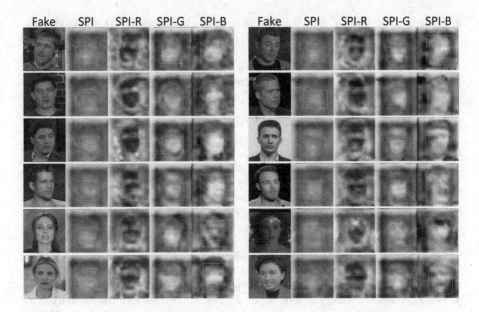

**Fig. 6.** Visual evaluation of our approach on Celeb-DF.

## 5   Conclusions

In this paper, we propose a novel framework named Find-X that aims to separate potentially inconsistent features of Deepfake videos. Unlike most traditional Deepfake detection methods that can only provide a probability value, Find-X attempts to visually explain the rationality of the detection results with a novel module SPI. SPI gives us another angle to analyze the authenticity of the video. In contrast to previous works, this is a new attempt to expose the nature of forgery. Extensive experiments demonstrate the efficiency and robustness of our method. Our method is useful for industrial applications and provides auxiliary evidence for victims.

**Acknowledgments.** This work was supported by National Key Technology Research and Development Program under 2020AAA0140000.

## References

1. Arnab, A., Dehghani, M., Heigold, G., Sun, C., Lučić, M., Schmid, C.: ViViT: a video vision transformer. In: 2021 IEEE/CVF International Conference on Computer Vision, ICCV 2021, Montreal, QC, Canada, pp. 6816–6826 (2021). https://doi.org/10.1109/ICCV48922.2021.00676
2. Chollet, F.: Xception: deep learning with depthwise separable convolutions. In: 2017 IEEE Conference on Computer Vision and Pattern Recognition, CVPR 2017, Honolulu, HI, USA, pp. 1800–1807 (2017). https://doi.org/10.1109/CVPR.2017.195

3. Cozzolino, D., Rössler, A., Thies, J., Nießner, M., Verdoliva, L.: ID-reveal: identity-aware deepfake video detection. In: 2021 IEEE/CVF International Conference on Computer Vision, ICCV 2021, Montreal, QC, Canada, pp. 15088–15097 (2021). https://doi.org/10.1109/ICCV48922.2021.01483

4. Diao, Q., Jiang, Y., Wen, B., Sun, J., Yuan, Z.: MetaFormer: a unified meta framework for fine-grained recognition. CoRR **abs/2203.02751** (2022). https://doi.org/10.48550/arXiv.2203.02751

5. Dufour, N., Gully, A.: DeepFakeDetection dataset (2019). https://ai.googleblog.com/2019/09/contributing-data-to-deepfake-detection.html

6. Fridrich, J.J., Kodovský, J.: Rich models for steganalysis of digital images. IEEE Trans. Inf. Forensics Secur. **7**(3), 868–882 (2012). https://doi.org/10.1109/TIFS.2012.2190402

7. Gu, Y., Zhao, X., Gong, C., Yi, X.: Deepfake video detection using audio-visual consistency. In: Zhao, X., Shi, Y.-Q., Piva, A., Kim, H.J. (eds.) IWDW 2020. LNCS, vol. 12617, pp. 168–180. Springer, Cham (2021). https://doi.org/10.1007/978-3-030-69449-4_13

8. Gu, Z., et al.: Spatiotemporal inconsistency learning for deepfake video detection. In: Shen, H.T., et al. (eds.) MM 2021: ACM Multimedia Conference, pp. 3473–3481. ACM, Virtual Event, China (2021). https://doi.org/10.1145/3474085.3475508

9. Gu, Z., Chen, Y., Yao, T., Ding, S., Li, J., Ma, L.: Delving into the local: dynamic inconsistency learning for deepfake video detection. In: Thirty-Sixth AAAI Conference on Artificial Intelligence, pp. 744–752. AAAI Press, Virtual Event (2022)

10. Guo, J., Han, K., Wu, H., Xu, C., Tang, Y., Xu, C., Wang, Y.: CMT: convolutional neural networks meet vision transformers. In: IEEE Conference on Computer Vision and Pattern Recognition, CVPR 2022, New Orleans, Louisiana (2022)

11. Haliassos, A., Vougioukas, K., Petridis, S., Pantic, M.: Lips don't lie: a generalisable and robust approach to face forgery detection. In: IEEE Conference on Computer Vision and Pattern Recognition, CVPR 2021, pp. 5039–5049. Virtual (2021)

12. Hu, J., Liao, X., Liang, J., Zhou, W., Qin, Z.: FInfer: frame inference-based deepfake detection for high-visual-quality videos. In: Thirty-Sixth AAAI Conference on Artificial Intelligence, pp. 951–959. AAAI Press, Virtual Event (2022)

13. Hu, Y., Zhao, H., Yu, Z., Liu, B., Yu, X.: Exposing deepfake videos with spatial, frequency and multi-scale temporal artifacts. In: Zhao, X., Piva, A., Comesaña-Alfaro, P. (eds.) IWDW 2021. LNCS, vol. 13180, pp. 47–57. Springer, Cham (2022). https://doi.org/10.1007/978-3-030-95398-0_4

14. Hu, Z., Xie, H., Wang, Y., Li, J., Wang, Z., Zhang, Y.: Dynamic inconsistency-aware deepfake video detection. In: Proceedings of the Thirtieth International Joint Conference on Artificial Intelligence, IJCAI 2021, pp. 736–742. Ijcai.org, Virtual Event/Montreal, Canada (2021). https://doi.org/10.24963/ijcai.2021/102

15. Jiang, Y., Chang, S., Wang, Z.: TransGAN: two pure transformers can make one strong GAN, and that can scale up. In: Advances in Neural Information Processing Systems 34: Annual Conference on Neural Information Processing Systems 2021, NeurIPS 2021, pp. 14745–14758. Virtual (2021)

16. Lee, C.C.: Elimination of redundant operations for a fast Sobel operator. IEEE Trans. Syst. Man Cybern. **13**(2), 242–245 (1983). https://doi.org/10.1109/TSMC.1983.6313122

17. Li, J., Xie, H., Li, J., Wang, Z., Zhang, Y.: Frequency-aware discriminative feature learning supervised by single-center loss for face forgery detection. In: IEEE Conference on Computer Vision and Pattern Recognition, CVPR 2021, pp. 6458–6467. Virtual (2021)

18. Li, L., et al.: Face X-ray for more general face forgery detection. In: 2020 IEEE/CVF Conference on Computer Vision and Pattern Recognition, CVPR 2020, Seattle, WA, USA, pp. 5000–5009 (2020). https://doi.org/10.1109/CVPR42600. 2020.00505

19. Li, Y., Yang, X., Sun, P., Qi, H., Lyu, S.: Celeb-DF: a large-scale challenging dataset for deepfake forensics. In: 2020 IEEE/CVF Conference on Computer Vision and Pattern Recognition, CVPR 2020, Seattle, WA, USA, pp. 3204–3213 (2020). https://doi.org/10.1109/CVPR42600.2020.00327

20. Liu, H., et al.: Spatial-phase shallow learning: Rethinking face forgery detection in frequency domain. In: IEEE Conference on Computer Vision and Pattern Recognition, CVPR 2021, pp. 772–781. Virtual (2021)

21. Liu, R., et al.: FuseFormer: fusing fine-grained information in transformers for video inpainting. In: 2021 IEEE/CVF International Conference on Computer Vision, ICCV 2021, Montreal, QC, Canada, pp. 14020–14029 (2021). https://doi.org/10. 1109/ICCV48922.2021.01378

22. Liu, Z., et al.: Swin transformer: hierarchical vision transformer using shifted windows. In: Proceedings of the IEEE/CVF International Conference on Computer Vision (ICCV) (2021)

23. Luo, Y., Zhang, Y., Yan, J., Liu, W.: Generalizing face forgery detection with high-frequency features. In: IEEE Conference on Computer Vision and Pattern Recognition, CVPR 2021, pp. 16317–16326. Computer Vision Foundation/IEEE, Virtual (2021). https://doi.org/10.1109/CVPR46437.2021.01605

24. Pei, P., Zhao, X., Li, J., Cao, Y., Yi, X.: Vision transformer based video hashing retrieval for tracing the source of fake videos. CoRR abs/2112.08117 (2021). https://arxiv.org/abs/2112.08117

25. Qian, Y., Yin, G., Sheng, L., Chen, Z., Shao, J.: Thinking in frequency: face forgery detection by mining frequency-aware clues. In: Vedaldi, A., Bischof, H., Brox, T., Frahm, J.-M. (eds.) ECCV 2020. LNCS, vol. 12357, pp. 86–103. Springer, Cham (2020). https://doi.org/10.1007/978-3-030-58610-2_6

26. Rössler, A., Cozzolino, D., Verdoliva, L., Riess, C., Thies, J., Nießner, M.: Face-Forensics++: learning to detect manipulated facial images. In: 2019 IEEE/CVF International Conference on Computer Vision, ICCV 2019, Seoul, Korea (South), pp. 1–11 (2019). https://doi.org/10.1109/ICCV.2019.00009

27. Sorkine, O., Cohen-Or, D., Lipman, Y., Alexa, M., Rössl, C., Seidel, H.: Laplacian surface editing. In: Boissonnat, J., Alliez, P. (eds.) Second Eurographics Symposium on Geometry Processing, Nice, France, 8–10 July 2004. ACM International Conference Proceeding Series, Nice, France, vol. 71, pp. 175–184 (2004). https:// doi.org/10.2312/SGP/SGP04/179-188

28. Sun, Z., Han, Y., Hua, Z., Ruan, N., Jia, W.: Improving the efficiency and robustness of deepfakes detection through precise geometric features. In: IEEE Conference on Computer Vision and Pattern Recognition, CVPR 2021, pp. 3609–3618. Virtual (2021)

29. Tan, M., Le, Q.V.: EfficientNet: rethinking model scaling for convolutional neural networks. In: Proceedings of the 36th International Conference on Machine Learning, ICML 2019, Long Beach, California, vol. 97, pp. 6105–6114 (2019)

30. Vaswani, A., et al.: Attention is all you need. In: Advances in Neural Information Processing Systems 30: Annual Conference on Neural Information Processing Systems 2017, Long Beach, CA, USA, pp. 5998–6008 (2017)

31. Wang, C., Deng, W.: Representative forgery mining for fake face detection. In: IEEE Conference on Computer Vision and Pattern Recognition, CVPR 2021, pp. 14923–14932. Virtual (2021)

32. Wang, W., Xie, E., Li, X., Fan, D.P., Shao, L.: PVTV 2: improved baselines with pyramid vision transformer. CoRR abs/2106.13797 (2021)

33. Wang, W., et al.: Pyramid vision transformer: a versatile backbone for dense prediction without convolutions. In: 2021 IEEE/CVF International Conference on Computer Vision, ICCV 2021, Montreal, QC, Canada, pp. 548–558 (2021). https://doi.org/10.1109/ICCV48922.2021.00061

34. Yang, C., Ma, J., Wang, S., Liew, A.W.: Preventing deepfake attacks on speaker authentication by dynamic lip movement analysis. IEEE Trans. Inf. Forensics Secur. **16**, 1841–1854 (2021). https://doi.org/10.1109/TIFS.2020.3045937

35. Yang, J., Li, A., Xiao, S., Lu, W., Gao, X.: MTD-net: learning to detect deepfakes images by multi-scale texture difference. IEEE Trans. Inf. Forensics Secur. **16**, 4234–4245 (2021). https://doi.org/10.1109/TIFS.2021.3102487

36. Yuan, Y., et al.: HRFormer: high-resolution vision transformer for dense predict. In: Advances in Neural Information Processing Systems 34: Annual Conference on Neural Information Processing Systems 2021, NeurIPS 2021, pp. 7281–7293. Virtual (2021)

37. Zhang, K., Zhang, Z., Li, Z., Qiao, Y.: Joint face detection and alignment using multitask cascaded convolutional networks. IEEE Signal Process. Lett. **23**(10), 1499–1503 (2016). https://doi.org/10.1109/LSP.2016.2603342

38. Zhao, H., Zhou, W., Chen, D., Wei, T., Zhang, W., Yu, N.: Multi-attentional deepfake detection. In: IEEE Conference on Computer Vision and Pattern Recognition, CVPR 2021, pp. 2185–2194. Virtual (2021)

39. Zhao, T., Xu, X., Xu, M., Ding, H., Xiong, Y., Xia, W.: Learning self-consistency for deepfake detection. In: 2021 IEEE/CVF International Conference on Computer Vision, ICCV 2021, Montreal, QC, Canada, pp. 15003–15013 (2021). https://doi.org/10.1109/ICCV48922.2021.01475

# Improving the Transferability of Adversarial Attacks Through Both Front and Rear Vector Method

Hao Wu[1] , Jinwei Wang[1,2]([⊠]), Jiawei Zhang[1], Xiangyang Luo[3], and Bin Ma[4]

[1] School of Computer and Software, Nanjing University of Information Science and Technology, Nanjing 210044, Jiangsu, China
wjwei_2004@163.com

[2] Engineering Research Center of Digital Forensics, Ministry of Education, Nanjing 210044, Jiangsu, China

[3] State Key Laboratory of Mathematical Engineering and Advanced Computing, Zhengzhou 450001, Henan, China
luoxy_ieu@sina.com

[4] Shandong Provincial Key Laboratory of Computer Networks and Qilu University of Technology, Jinan 250353, Shandong, China

**Abstract.** Deep Neural Networks (DNNs) are vulnerable to adversarial attacks, which makes adversarial attacks serve as a method to evaluate the robustness of DNNs. However, adversarial attacks have the disadvantage of high white-box attack success rates but low transferability. Therefore, many methods were proposed to improve the transferability of adversarial attacks, one of which is the momentum-based method. To improve the transferability of the existing adversarial attacks, we propose Previous-gradient as Neighborhood NI-FGSM (PN-NI-FGSM) and Momentum as Neighborhood NI-FGSM (MN-NI-FGSM), both of which are the momentum-based attacks. The results show that momentum describes the neighborhood more preciselfy than the previous gradient. Additionally, we define the front vector and the rear vector. Then, we classify momentum-based attacks into front vector attacks and rear vector attacks. Finally, we propose Both Front and Rear Vector Method (BFRVM), which combines the front vector attacks and the rear vector attacks. The experiments show that our BFRVM attacks achieve the best transferability against normally trained models and adversarially trained models under the single-model setting and ensemble-model setting, respectively.

**Keywords:** Adversarial attacks · Transferability · Momentum

## 1 Introduction

Deep Neural Networks (DNNs) [11,13,14,24,31] have been widely applied in computer vision, such as autonomous driving [3,6,10], and facial recognition [2,7]. However, Szegedy [26] found that applying certain imperceptible perturbations to images can make DNNs misclassify, and they refer to such perturbed images as adversarial examples. Adversarial examples raise security concerns for DNNs.

Gradient-based attacks [8,16,19] are widely studied because they have low time cost. Gradient-based attacks can be classified into one-step attacks [8] and iterative attacks [4,16]. Generally, iterative attacks have higher white-box attack success rates but lower transferability than one-step attacks, which makes iterative attacks difficult in black-box settings. Therefore, improving the transferability of gradient-based iterative attacks [1,4,16] has become a research hotspot.

Many methods [4,5,16,18,28,30] have been proposed to improve the transferability of gradient-based iterative attacks. Dong [4] proposed MI-FGSM, which integrates momentum into I-FGSM. NI-FGSM [18] integrates Nesterov Accelerated Gradient (NAG) [20] into I-FGSM. Input transformations (e.g., DIM [30], TIM [5] and SIM [18]), which transform inputs before they are fed into DNNs, were proposed to improve transferability. EMI-FGSM and ENI-FGSM [28], which use the previous gradient and momentum to describe the neighborhood, respectively, replace the gradient at the current data point with the average gradient within the neighborhood of the current data point to improve MI-FGSM. Wang [28] argued that momentum cannot be used as a precise description of the neighborhood because results show that the transferability of ENI-FGSM is lower than EMI-FGSM. However, we consider momentum as the accumulation of previous gradients to be more robust than the previous gradient and demonstrate that momentum can more accurately describe the neighborhood.

We also use the same methods as EMI-FGSM and ENI-FGSM to improve the transferability of NI-FGSM. Specifically, we propose Previous-gradient as Neighborhood NI-FGSM (PN-NI-FGSM) and Momentum as Neighborhood NI-FGSM (MN-NI-FGSM) that use the previous gradient and momentum to describe the neighborhood of the pre-update point, respectively.

In this paper, we define the front vector and the rear vector (Definition 1 in Sect. 3), and then we classify existing gradient-based momentum iterative attacks into front vector attacks and rear vector attacks. During each iteration, front vector attacks (e.g., MI-FGSM, EMI-FGSM, and ENI-FGSM) update along a certain gradient before updating along momentum. Such gradient is the front vector, and the rear vector is the zero vector. Similarly, rear vector attacks (e.g., NI-FGSM, PN-NI-FGSM, and MN-NI-FGSM) modify the pre-update along a certain gradient after pre-updating along momentum. Such gradient is the rear vector, and the front vector is the zero vector. However, the rear vector attacks pre-update along momentum, which is simple and rough. Therefore, we refine the pre-update by adding the front vector into each pre-update, which we call Both Front and Rear Vector Method (BFRVM). Since we can combine a certain front vector attack and a certain rear vector attack to obtain a BFRVM attack, we call the BFRVM attack child attack, and attacks that derive the BFRVM attack are called parent attacks. Some parent attacks and BFRVM child attacks are shown in Table 1. Overall, we make the following contributions:

**Table 1.** Parent attacks and BFRVM child attacks

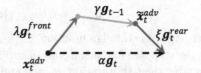

**Fig. 1.** Illustration of BFRVM: $g_t^{front}$ is the front vector, $g_t^{rear}$ is the rear vector, $\gamma g_{t-1}$ is the momentum term, $g_t$ is the updated momentum, and $\widetilde{x}_t^{adv}$ is the pre-update point

| Parent attacks | | Child attacks |
|---|---|---|
| Front vector attacks | Rear vector attacks | |
| MI-FGSM | NI-FGSM | BI-FGSM (Ours) |
| EMI-FGSM | PN-NI-FGSM (Ours) | PN-BI-FGSM (Ours) |
| ENI-FGSM | MN-NI-FGSM (Ours) | MN-BI-FGSM (Ours) |

- We demonstrate that momentum describes the neighborhood more precisely and robustly than the previous gradient.
- We propose PN-NI-FGSM and MN-NI-FGSM, which use the previous gradient and momentum to describe the neighborhood of the pre-update point, respectively, to improve the transferability of NI-FGSM.
- We define the front vector and the rear vector, and we classify existing gradient-based momentum iterative attacks into front vector attacks and rear vector attacks. Additionally, we propose BFRVM attacks to improve transferability by combining front vector attacks and rear vector attacks.

## 2   Related Work

In this section, we introduce gradient-based iterative attacks, input transformations, ensemble in logits, and adversarial training. Given a classifier $f(x; \theta)$, where $x$ is an input, and $\theta$ is parameters of $f$. Let $J(\cdot, y)$ is a loss function, where $y$ is the true label of the input $x$. An adversarial example $x^{adv}$ satisfies $f(x; \theta) \neq f(x^{adv}; \theta)$ under the constraint of $\left\| x^{adv} - x \right\|_p \leq \epsilon$, where $\|\cdot\|_p$ denotes the $L_p$ norm, and $p$ is generally 0, 1, 2, $\infty$. We focus on $p = \infty$.

### 2.1   Adversarial Attack Methods

We introduce three methods to improve the transferability of gradient-based iterative attacks: momentum-based methods, input transformations, and ensemble in logits.

**Momentum-Based Methods.** Momentum dampens gradient oscillation to accelerate gradient ascent. We introduce the gradient-based momentum iterative attacks here. Momentum I-FGSM (MI-FGSM) [4] integrates the momentum method [21] into I-FGSM [16] to accelerate gradient ascent:

$$g_t = \mu \cdot g_{t-1} + \frac{\nabla_{x_t^{adv}} J(f(x_t^{adv}; \theta), y))}{\left\| \nabla_{x_t^{adv}} J(f(x_t^{adv}; \theta), y)) \right\|_1}, \tag{1}$$

$$x_{t+1}^{adv} = Clip_{(x,\epsilon)}\{x_t^{adv} + \alpha \cdot \text{sign}(g_t)\}, \tag{2}$$

where $\text{sign}(\cdot)$ is the sign function, $Clip_{(x,\epsilon)}\{x'\}$ denotes $\min(\max(x', x-\epsilon), x+\epsilon)$, $\alpha$ is the step size for each iteration, $x_1^{adv} = x, t = 1, 2, \cdots, T, g_t$ denotes momentum after the $t$-th iteration, $g_0 = 0$ and $\mu$ is the decay factor. Similarly, Nesterov I-FGSM (NI-FGSM) [18] integrates Nesterov Accelerated Gradient (NAG) [20] into I-FGSM. Unlike MI-FGSM, NI-FGSM provides the pre-update by replacing $x_t^{adv}$ with $x_t^{adv} + \alpha \cdot \mu \cdot g_{t-1}$ in Eq. 1. Enhanced MI-FGSM (EMI-FGSM) [28] replaces the gradient of the current data point with the average gradient within the neighborhood of the current data point to improve the transferability of MI-FGSM. EMI-FGSM uses the previous gradient to describe the neighborhood:

$$\bar{x}_t^{adv}[i] = x_t^{adv} + c_i \cdot \bar{g}_{t-1}, \tag{3}$$

$$\bar{g}_t = \frac{1}{N} \sum_{i=1}^{N} \nabla_{\bar{x}_t^{adv}[i]} J(f(\bar{x}_t^{adv}[i]; \theta), y)), \tag{4}$$

$$g_t = \mu \cdot g_{t-1} + \frac{\bar{g}_t}{\|\bar{g}_t\|_1}, \tag{5}$$

$$x_{t+1}^{adv} = Clip_{(x,\epsilon)}\{x_t^{adv} + \alpha \cdot \text{sign}(g_t)\}, \tag{6}$$

where $\bar{g}_0 = 0$, $c_i \in [-\eta, \eta]$ denotes the $i$-th sampling coefficient, and $N$ denotes the number of sampling examples. Unlike EMI-FGSM, Enhanced NI-FGSM (ENI-FGSM) [28] uses momentum to describe the neighborhood of the current data point to improve the transferability of MI-FGSM. $\bar{g}_{t-1}$ is replaced by the momentum $g_{t-1}$ in Eq. 3.

**Input Transformations.** Input transformations transform the input before it is fed into the DNN to improve transferability. Diverse Input Method (DIM) [30] performs random resizing and padding on the input with probability $p$ before feeding the input into the DNN. Translation-Invariant Method (TIM) [5] convolves the gradient to approximate the translation of the input. TIM especially shows better transferability against models with defense [9,17,29]. Scale-Invariant Method (SIM) [18] scales the input with the scale factor $1/2^i$ to derive an ensemble of models. SIM is an approximate method of ensemble in logits [4].

**Ensemble in Logits.** Ensemble in logits [4] generates adversarial examples on an ensemble of models whose logit activations are fused together. Specifically, the logits of $K$ models are fused as $l(x) = \sum_{k=1}^{K} w_k l_k(x)$, where $l_k(x)$ denotes the logits of the $k$-th model, $w_k$ is the ensemble weight with $w_k \geq 0$ and $\sum_{k=1}^{K} w_k = 1$. Adversarial examples generated on the ensemble model have higher transferability.

## 2.2   Adversarial Training

Adversarial training increases robustness by adding adversarial examples to the training data. Goodfellow [8] showed that adversarially trained models are more robust. However, Kurakin [16] pointed out that adversarial training is not robust to iterative attacks. What's more, Tramèr [27] showed that adversarially trained models are still vulnerable to simple white-box and black-box attacks. Therefore, they proposed ensemble adversarial training, which adds adversarial examples generated from other models to the training data, to increase robustness.

## 3   Methodology

In this section, we focus on improving the transferability of gradient-based momentum iterative attacks. We first propose PN-NI-FGSM and MN-NI-FGSM to improve NI-FGSM. What's more, we propose a gradient-based momentum iterative method involving the front vector and the rear vector, called Both Front and Rear Vector Method (BFRVM). Finally, we propose BFRVM I-FGSM (BI-FGSM), Previous-Gradient as Neighborhood BFRVM I-FGSM (PN-BI-FGSM) and Momentum as Neighborhood BFRVM I-FGSM (MN-BI-FGSM) by integrating BFRVM into concrete attacks. We define the front vector and the rear vector here:

**Definition 1.** *The front vector and the rear vector. Given a $C \times H \times W$ dimensional vector space $\boldsymbol{X} = \{\boldsymbol{x}|\boldsymbol{x} = (x_{ijk})_{C \times H \times W} \in \mathbb{R}^{C \times H \times W}, i = 1, 2, \cdots, C, j = 1, 2, \cdots, H, k = 1, 2, \cdots, W\}$ on $\mathbb{R}$, and a momentum iteration in the vector space $\boldsymbol{X}$:*

$$\begin{cases} \boldsymbol{g}_0 = \boldsymbol{0} \in \boldsymbol{X}, \\ \boldsymbol{g}_t = a \cdot \boldsymbol{g}_t^{front} + b \cdot \boldsymbol{g}_{t-1} + c \cdot \boldsymbol{g}_t^{rear}, \end{cases} \quad (7)$$

*where $\alpha, a, b, c \in \mathbb{R}$, $\boldsymbol{g}_t^{front}, \boldsymbol{g}_{t-1}, \boldsymbol{g}_t^{rear}, \boldsymbol{g}_t \in \boldsymbol{X}$, $t = 1, 2, \cdots, T$, and $b \cdot \boldsymbol{g}_{t-1}$ is the momentum term. If there is a logical relation that vectors $\boldsymbol{g}_t^{front}$, $\boldsymbol{g}_{t-1}$, and $\boldsymbol{g}_t^{rear}$ connected end to end, we call $\boldsymbol{g}_t^{front}$ the front vector and $\boldsymbol{g}_t^{rear}$ the rear vector.*

### 3.1   Previous-Gradient and Momentum as Neighborhood NI-FGSM

As shown in Fig. 2b, 2c, EMI-FGSM and ENI-FGSM use the average gradient within the neighborhood of the current data point instead of the gradient at the current data point as in MI-FGSM (see Fig. 2a). Specifically, EMI-FGSM and ENI-FGSM use the previous gradient and momentum to describe the neighborhood of the current data point, respectively. Then they sample along the direction of the previous gradient and momentum near the current data point, respectively, and use the average gradient of the sampled examples as the average gradient within the neighborhood approximately. Similarly, as shown in

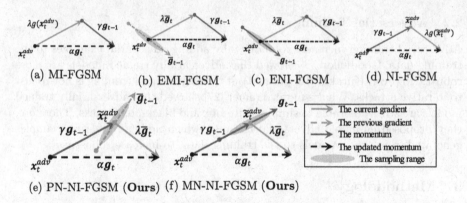

(a) MI-FGSM     (b) EMI-FGSM     (c) ENI-FGSM     (d) NI-FGSM

(e) PN-NI-FGSM (**Ours**) (f) MN-NI-FGSM (**Ours**)

**Fig. 2.** Illustration of front vector attacks and rear vector attacks: $g(\cdot)$ denotes the function for calculating the gradient, (a), (b), (c) belong to front vector attacks, (d), (e), (f) belong to rear vector attacks.

Fig. 2e, 2f, we also can use the average gradient within the neighborhood of the pre-update point instead of the gradient at the pre-update point as in NI-FGSM (see Fig. 2d). Therefore, we propose PN-NI-FGSM which uses the previous gradient to describe the neighborhood of the pre-update point and MN-NI-FGSM which uses momentum to describe the neighborhood of the pre-update point. The following is the specific description of PN-NI-FGSM and MN-NI-FGSM:

**PN-NI-FGSM.** As shown in Fig. 2e, during each iteration, PN-NI-FGSM first pre-updates along the direction of momentum $g_{t-1}$ to obtain the pre-update point $\widetilde{x}_t^{adv}$:

$$\widetilde{x}_t^{adv} = x_t^{adv} + \alpha \cdot \mu \cdot g_{t-1}, \tag{8}$$

where $\alpha$ is the step size for each iteration, and $\mu$ is the decay factor. Next, PN-NI-FGSM uses the previous gradient to describe the neighborhood of the pre-update point, and samples along the direction of the previous gradient near the pre-update point. Then PN-NI-FGSM calculates the average gradient of the sampled examples:

$$\bar{x}_t^{adv}[i] = \widetilde{x}_t^{adv} + c_i \cdot \bar{g}_{t-1}, \tag{9}$$

$$\bar{g}_t = \frac{1}{N} \sum_{i=1}^{N} \nabla_{\bar{x}_t^{adv}[i]} J(f(\bar{x}_t^{adv}[i]; \theta), y)), \tag{10}$$

where $\bar{g}_t$ denotes the average gradient during the $t$-th iteration, $\bar{g}_0 = 0$, $c_i \in [-\eta, \eta]$ denotes the $i$-th sampling coefficient, and $N$ denotes the number of sampling examples. Finally, PN-NI-FGSM updates along the direction of the average gradient to modify the pre-update:

$$g_t = \mu \cdot g_{t-1} + \frac{\bar{g}_t}{\|\bar{g}_t\|_1}, \tag{11}$$

$$x_{t+1}^{adv} = Clip_{(x,\epsilon)}\{x_t^{adv} + \alpha \cdot \text{sign}(g_t)\}. \tag{12}$$

---

**Algorithm 1:** MN-BI-FGSM

---

    **Input**  : A classifier $f$ and a loss function $J$. A clean example $x$. and its real
                   label $y$.

    **Input**  : The maximum perturbation $\epsilon$, number of iteration $T$ and decay factor
                   $\mu$.

    **Input**  : The sampling interval bound $\eta$ and the sampling number $N$.

    **Output:** An adversarial example $x^{adv}$.

1  $\alpha \leftarrow \epsilon/T$; $g_0 \leftarrow 0$; $x_1^{adv} \leftarrow x$;

2  **for** $t \leftarrow 1$ **to** $T$ **do**

3      $\bar{x}_t^{adv}[i] \leftarrow x_t^{adv} + c_i \cdot g_{t-1}$, $c_i \in [-\eta, \eta]$ ;          // Sampling

4      $\bar{g}_t \leftarrow \frac{1}{N}\sum_{i=1}^{N} \nabla_{\bar{x}_t^{adv}[i]} J(f(\bar{x}_t^{adv}[i]; \theta), y))$ ;    // The average gradient
      serves as the front vector

5      $\widetilde{x}_t^{adv} \leftarrow x_t^{adv} + \frac{\bar{g}_t}{\|\bar{g}_t\|_1} + \alpha \cdot \mu \cdot g_{t-1}$ ;        // Pre-update

6      $\hat{x}_t^{adv}[i] \leftarrow \widetilde{x}_t^{adv} + c_i \cdot g_{t-1}$, $c_i \in [-\eta, \eta]$ ;       // Sampling

7      $\hat{g}_t \leftarrow \frac{1}{N}\sum_{i=1}^{N} \nabla_{\hat{x}_t^{adv}[i]} J(f(\hat{x}_t^{adv}[i]; \theta), y))$ ;    // The average gradient
      serves as the rear vector

8      $g_t \leftarrow \frac{\bar{g}_t}{\|\bar{g}_t\|_1} + \mu \cdot g_{t-1} + \frac{\hat{g}_t}{\|\hat{g}_t\|_1}$ ;         // Update momentum

9      $x_{t+1}^{adv} \leftarrow Clip_{(x,\epsilon)}\{x_t^{adv} + \alpha \cdot \text{sign}(g_t)\}$ ;   // Update adversarial example

10 **end**

11 **return** $x^{adv} \leftarrow x_{T+1}^{adv}$.

---

**MN-NI-FGSM.** As shown in Fig. 2f, during each iteration, MN-NI-FGSM differs from PN-NI-FGSM in that MN-NI-FGSM uses momentum to describe the neighborhood of the pre-update point, not the previous gradient. Therefore, Eq. 9 is replaced with the following equation:

$$\bar{x}_t^{adv}[i] = x_t^{adv} + c_i \cdot g_{t-1}. \tag{13}$$

We study the hyperparameters $\eta$ and $N$ of MN-NI-FGSM in Sect. 4.4, and we compare the transferability of PN-NI-FGSM and MN-NI-FGSM in Sect. 4.2. The results show that momentum is more robust than the previous gradient and describes the neighborhood more precisely.

### 3.2  Both Front and Rear Vector Method

As shown in Fig. 2, we classify existing gradient-based momentum iterative attacks into front vector attacks and rear vector attacks. During each iteration, Front vector attacks (e.g., MI-FGSM, EMI-FGSM, and ENI-FGSM) update along the front vector before updating along momentum. Rear vector attacks (e.g., NI-FGSM, PN-NI-FGSM, and MN-NI-FGSM) pre-update along momentum and then modify the pre-update along the rear vector. However, pre-updating only along momentum is rough. Therefore, We add the front vector into the pre-update to refine the pre-update. We call the momentum iterative method with the front vector and the rear vector BFRVM.

As shown in Fig. 1, during each iteration, the pre-update of BFRVM is performed successively along the direction of the front vector $g_t^{front}$ and momentum $g_{t-1}$, and then the pre-update is modified along the rear vector $g_t^{rear}$. Therefore, as shown in Table 1, we can get the BFRVM attack (i.e., child attack) by combining one front vector attack and one rear vector attack (i.e., parent attack). The algorithm of MN-BI-FGSM is summarized in Algorithm 1. The algorithms of BI-FGSM and PN-BI-FGSM are similar to Algorithm 1.

## 4   Experiments

### 4.1   Setup

**Models:** Four normal trained models, i.e., Inception-v3 (Inc-v3) [25], Inception-v4 (Inc-v4) [23], Inception-Resnet-v2 (IncRes-v2) [23] and Resnet-152 (Res-152) [12], one normal adversarially trained model, i.e., adv-Inception-v3 (Inc-v3_adv) [15], and one ensemble adversarially trained model, i.e., ens-adv-Inception-Resnet-v2 (IncRes-v2_ens) [27].

**Dataset:** We randomly select 1000 images from the ILSVRC 2012 validation set [22], all of which can be correctly classified by the above models. To ensure that all images are the same size, we first scale these images to a height of 256 while maintaining the aspect ratio, and then crop the images to $224 \times 224$ at the center.

**Baselines:** We compare our methods with FGSM [8], I-FGSM, MI-FGSM, NI-FGSM, EMI-FGSM, and ENI-FGSM. In addition, we integrate input transformations (e.g., DIM, TIM, SIM, and DTS, which is the integration of the former three) into MN-BI-FGSM, which is our best method, and compare them with input transformations.

**Hyperparameters:** We set the maximum perturbation $\epsilon = {}^{16}/_{255}$ for all attacks. For iterative attacks, we set the number of iterations $T = 10$, and the iterative step size $\alpha = \epsilon/T$. For momentum-based iterative attacks, we set the decay factor $\mu = 1.0$ as in [4,18]. For DIM, we set the transformation probability $p = 0.5$. For TIM, we set the size of the Gaussian kernel to $15 \times 15$. For SIM, we set the number of scale copies $m = 5$. For EMI-FGSM and PN-NI-FGSM, we set the sampling interval bound $\eta = 7$. For ENI-FGSM and MN-NI-FGSM, we set the sampling interval bound $\eta = 0.02$. For all attacks with sampling, we set the number of sampling examples $N = 11$. We adopt the linear sampling, which is the same as [28]. For the hyperparameters of all BFRVM child attacks, we set the same values as their parent attacks. For attacking ensemble of models, the ensemble weight of all ensembled models is the same, i.e., $w_k = {}^1/_5$.

### 4.2   Attack Single Models

We generate adversarial examples on single-models and compare attack success rates. We also integrate input transformations into our best BFRVM attacks to test attack success rates. All attacks are untargeted attacks based on $L_\infty$ norm bound, and the hyperparameters are set as in Sect. 4.1 and Sect. 4.4.

**Table 2.** The attack success rates (%) of adversarial attacks generated on Inc-v3, Inc-v4, IncRes-v3, and Res-152, respectively, against the six baseline models under the single-model setting, "*" indicates the model being white-box attacked

| Models | Attacks | Inc-v3 | Inc-v4 | IncRes-v2 | Res-152 | Inc-v3_adv | IncRes-v2_ens |
|---|---|---|---|---|---|---|---|
| Inc-v3 | MI-FGSM | 98.9* | 42.7 | 37.0 | 27.6 | 29.2 | 12.7 |
| | NI-FGSM | 99.3* | 53.8 | 46.4 | 37.9 | 29.6 | 14.5 |
| | BI-FGSM (Ours) | 99.2* | **56.4** | **49.7** | **40.1** | **34.4** | **14.7** |
| | EMI-FGSM | 98.8* | 53.1 | 48.2 | 37.7 | 30.6 | 15.0 |
| | PN-NI-FGSM (Ours) | 99.2* | 57.9 | 51.1 | 40.5 | 31.0 | 14.8 |
| | PN-BI-FGSM (Ours) | 98.8* | **61.9** | **55.8** | **44.2** | **36.0** | 14.9 |
| | ENI-FGSM | 99.8* | 62.9 | 56.2 | 46.2 | 34.1 | 17.5 |
| | MN-NI-FGSM (Ours) | 99.9* | 65.9 | 56.9 | 48.1 | 34.0 | 16.7 |
| | MN-BI-FGSM (Ours) | 99.9* | **66.1** | **58.6** | **48.9** | **36.2** | **18.7** |
| Inc-v4 | MI-FGSM | 42.0 | 98.1* | 36.1 | 25.6 | 28.4 | 12.7 |
| | NI-FGSM | 45.4 | 99.1* | 39.2 | 28.1 | 28.9 | 13.0 |
| | BI-FGSM (Ours) | **50.5** | 99.0* | **44.6** | **32.4** | **35.3** | **14.3** |
| | EMI-FGSM | 48.6 | 98.1* | 43.4 | 29.7 | 31.3 | 14.9 |
| | PN-NI-FGSM (Ours) | 48.4 | **99.1*** | 41.6 | 29.7 | 29.9 | 13.6 |
| | PN-BI-FGSM (Ours) | **55.6** | 98.8* | **47.8** | **35.5** | **36.4** | 14.5 |
| | ENI-FGSM | 53.2 | 99.8* | 48.4 | 34.3 | 32.8 | 15.3 |
| | MN-NI-FGSM (Ours) | 55.7 | **100.0*** | 51.1 | 34.2 | 34.1 | 15.9 |
| | MN-BI-FGSM (Ours) | **58.6** | **100.0*** | **52.5** | **36.2** | 34.9 | **16.2** |
| IncRes-v2 | MI-FGSM | 44.1 | 47.1 | 98.5* | 27.3 | 33.2 | 20.0 |
| | NI-FGSM | 45.3 | 48.6 | **99.0*** | 30.3 | 33.5 | 19.6 |
| | BI-FGSM (Ours) | **53.5** | **55.5** | 98.1* | **35.0** | **37.8** | **21.1** |
| | EMI-FGSM | 51.1 | 52.1 | 98.5* | 32.9 | 36.5 | 22.3 |
| | PN-NI-FGSM (Ours) | 49.0 | 50.8 | **99.3*** | 31.3 | 34.1 | 21.7 |
| | PN-BI-FGSM (Ours) | **57.7** | **59.3** | 98.1* | **37.0** | **39.9** | **22.5** |
| | ENI-FGSM | 56.1 | 56.8 | 99.7* | 37.4 | 38.3 | 24.7 |
| | MN-NI-FGSM (Ours) | 59.0 | 60.7 | **100.0*** | 38.2 | 38.5 | 25.8 |
| | MN-BI-FGSM (Ours) | **60.5** | **62.8** | **100.0*** | **38.9** | 39.3 | **27.4** |
| Res-152 | MI-FGSM | 48.2 | 41.5 | 34.0 | 99.9* | 29.0 | 14.4 |
| | NI-FGSM | 51.4 | 45.9 | 37.4 | 99.9* | 28.5 | 15.1 |
| | BI-FGSM (Ours) | **58.2** | **52.2** | 45.0 | 99.9* | **34.2** | 15.6 |
| | EMI-FGSM | 53.8 | 48.2 | 40.2 | 99.9* | 30.0 | 15.3 |
| | PN-NI-FGSM (Ours) | 55.2 | 47.9 | 40.1 | **100.0*** | 29.0 | 16.5 |
| | PN-BI-FGSM (Ours) | **62.0** | **54.7** | **46.9** | 99.9* | **35.2** | 16.6 |
| | ENI-FGSM | 62.8 | 55.9 | 46.1 | **100.0*** | 33.2 | 17.9 |
| | MN-NI-FGSM (Ours) | 63.2 | 58.5 | 47.4 | **100.0*** | 32.7 | 17.8 |
| | MN-BI-FGSM (Ours) | **67.8** | **61.5** | **51.8** | **100.0*** | 32.8 | **19.5** |

**Parent Attacks vs. Child Attacks.** We compare the attack success rates of BFRVM child attacks (i.e., BI-FGSM, PN-BI-FGSM, and MN-BI-FGSM) and their parent attacks against the six baseline models under the single-model setting. The attack success rates of adversarial examples generated on Inc-v3, Inc-v4, IncRes-v3, and Res-152, respectively, are shown in Table 2. Next, we take the adversarial samples generated on Inc-v3 as an example for specific

**Table 3.** The attack success rates (%) of adversarial attacks integrated with input transformations against the six baseline models under the single-model setting, The adversarial examples are generated on Inc-v3, Inc-v4, IncRes-v3, and Res-152, respectively. "*" indicates the model being white-box attacked

| Models | Attacks | Inc-v3 | Inc-v4 | IncRes-v2 | Res-152 | Inc-v3$_{adv}$ | IncRes-v2$_{ens}$ |
|---|---|---|---|---|---|---|---|
| Inc-v3 | MN-BI-FGSM (Ours) | 99.9* | 66.1 | 58.6 | 48.9 | 36.2 | 18.7 |
|  | MN-BI-DIM (Ours) | 99.9* | 78.2 | 73.8 | 61.9 | 46.2 | 28.8 |
|  | MN-BI-FGSM (Ours) | 99.9* | 66.1 | 58.6 | 48.9 | 36.2 | 18.7 |
|  | MN-BI-TIM (Ours) | 99.9* | 68.2 | 59.9 | 54.3 | 41.8 | 28.8 |
|  | MN-BI-FGSM (Ours) | 99.9* | 66.1 | 58.6 | 48.9 | 36.2 | 18.7 |
|  | MN-BI-SIM (Ours) | 100.0* | 80.6 | 76.1 | 64.5 | 50.9 | 32.3 |
|  | MN-BI-FGSM (Ours) | 99.9* | 66.1 | 58.6 | 48.9 | 36.2 | 18.7 |
|  | MN-BI-DTS (Ours) | 100.0* | 86.3 | 84.0 | 64.0 | 67.8 | 57.9 |
| Inc-v4 | MN-BI-FGSM (Ours) | 58.6 | 100.0* | 52.5 | 36.2 | 34.9 | 16.2 |
|  | MN-BI-DIM (Ours) | 75.4 | 99.8* | 70.5 | 52.2 | 42.9 | 24.7 |
|  | MN-BI-FGSM (Ours) | 58.6 | 100.0* | 52.5 | 36.2 | 34.9 | 16.2 |
|  | MN-BI-TIM (Ours) | 61.5 | 99.9* | 55.6 | 39.5 | 41.0 | 27.5 |
|  | MN-BI-FGSM (Ours) | 58.6 | 100.0* | 52.5 | 36.2 | 34.9 | 16.2 |
|  | MN-BI-SIM (Ours) | 80.7 | 100.0* | 73.8 | 53.2 | 43.1 | 23.2 |
|  | MN-BI-FGSM (Ours) | 58.6 | 100.0* | 52.5 | 36.2 | 34.9 | 16.2 |
|  | MN-BI-DTS (Ours) | 90.3 | 100.0* | 87.0 | 63.1 | 61.0 | 51.9 |
| IncRes-v2 | MN-BI-FGSM (Ours) | 60.5 | 62.8 | 100.0* | 38.9 | 39.3 | 27.4 |
|  | MN-BI-DIM (Ours) | 75.9 | 76.0 | 99.9* | 53.8 | 49.3 | 39.0 |
|  | MN-BI-FGSM (Ours) | 60.5 | 62.8 | 100.0* | 38.9 | 39.3 | 27.4 |
|  | MN-BI-TIM (Ours) | 65.2 | 65.3 | 100.0* | 43.6 | 48.3 | 40.7 |
|  | MN-BI-FGSM (Ours) | 60.5 | 62.8 | 100.0* | 38.9 | 39.3 | 27.4 |
|  | MN-BI-SIM (Ours) | 71.8 | 68.2 | 99.9* | 44.2 | 48.1 | 32.3 |
|  | MN-BI-FGSM (Ours) | 60.5 | 62.8 | 100.0* | 38.9 | 39.3 | 27.4 |
|  | MN-BI-DTS (Ours) | 83.6 | 81.5 | 99.8* | 55.6 | 66.1 | 60.5 |
| Res-152 | MN-BI-FGSM (Ours) | 67.8 | 61.5 | 51.8 | 100.0* | 32.8 | 19.5 |
|  | MN-BI-DIM (Ours) | 87.5 | 83.7 | 75.1 | 100.0* | 43.8 | 30.2 |
|  | MN-BI-FGSM (Ours) | 67.8 | 61.5 | 51.8 | 100.0* | 32.8 | 19.5 |
|  | MN-BI-TIM (Ours) | 69.1 | 63.3 | 57.5 | 100.0* | 43.1 | 31.4 |
|  | MN-BI-FGSM (Ours) | 67.8 | 61.5 | 51.8 | 100.0* | 32.8 | 19.5 |
|  | MN-BI-SIM (Ours) | 83.6 | 76.9 | 69.9 | 100.0* | 43.9 | 30.1 |
|  | MN-BI-FGSM (Ours) | 67.8 | 61.5 | 51.8 | 100.0 | 32.8 | 19.5 |
|  | MN-BI-DTS (Ours) | 93.0 | 89.1 | 87.1 | 100.0* | 68.5 | 66.3 |

analysis. For front vector attacks, the black-box attack success rates of EMI-FGSM and ENI-FGSM are ~7.1% and ~13.5% higher than MI-FGSM on average, respectively, which indicates that the average gradient of the neighborhood is more precise than the current gradient. Meanwhile, the black-box attack success rates of ENI-FGSM are ~6.5% higher than EMI-FGSM on average, which indicates that momentum describes the neighborhood more precisely than the previous gradient. Similar conclusions can be drawn for rear vector attacks (i.e., NI-FGSM, PN-NI-FGSM, and MN-NI-FGSM) and BFRVM attacks (i.e., BI-FGSM, PN-BI-FGSM, and MN-BI-FGSM). Additionally, the black-box attack

**Table 4.** The attack success rates (%) of adversarial attacks against the six baseline models under the ensemble-model setting. The ensemble models test white-box attack success rates and the hold-out models test black-box attack success rates, "-" indicates the hold-out model

| Model | Attacks | -Inc-v3 | -Inc-v4 | -IncRes-v2 | -Res-152 | -Inc-v3$_{adv}$ | -IncRes-v2$_{ens}$ |
|---|---|---|---|---|---|---|---|
| Ensemble | FGSM | 40.8 | 40.8 | 38.7 | 43.4 | 52.7 | 53.3 |
| | I-FGSM | 99.5 | 99.4 | 99.5 | 99.5 | **99.9** | 99.8 |
| | ENI-FGSM | 99.9 | **99.9** | **99.9** | **99.9** | **99.9** | **99.9** |
| | MN-NI-FGSM (Ours) | 99.9 | **99.9** | **99.9** | **99.9** | **99.9** | **99.9** |
| | MN-BI-FGSM (Ours) | **100.0** | **99.9** | **99.9** | **99.9** | **99.9** | **99.9** |
| Hold-out | FGSM | 40.2 | 39.5 | 37.8 | 29.5 | 39.1 | 16.5 |
| | I-FGSM | 26.7 | 32.9 | 26.6 | 13.8 | 25.9 | 11.1 |
| | ENI-FGSM | 70.5 | 72.7 | 66.3 | 53.1 | 46.2 | 23.7 |
| | MN-NI-FGSM (Ours) | 70.3 | 74.9 | 67.6 | 54.5 | 47.0 | 25.0 |
| | MN-BI-FGSM (Ours) | **73.4** | **77.6** | **71.1** | **55.8** | **47.2** | **25.5** |

success rates of three BFRVM attacks (i.e., BI-FGSM, PN-BI-FGSM, and MN-BI-FGSM) are ~5.9%, ~4.6%, and ~1.9% higher than their parent attacks on average, respectively. However, our methods have a limitation, which is that when improving transferability, may reduce the white-box attacks success rate. Our MN-BI-FGSM shows the best performance. The white-box attack success rate of MN-BI-FGSM is 99.9% and the black-box attack success rates are ~4.9% higher than other BFRVM attacks (i.e., BI-FGSM and PN-BI-FGSM) on average.

**Integrated with Input Transformations.** Since our MN-BI-FGSM shows the best attack success rates, we integrate input transformations (i.e., DIM, TIM, SIM, and DTS) into MN-BI-FGSM to further improve transferability. We compare the attack success rates of MN-BI-FGSM and MN-BI-FGSM integrated with input transformations in Table 3. Adversarial examples are generated on Inc-v3, Inc-v4, IncRes-v3, and Res-152, respectively. Next, we take the adversarial samples generated on Inc-v3 as an example for specific analysis. The white-box attack success rates of MN-BI-FGSM integrated with input transformations are not lower than MN-BI-FGSM. The black-box attack success rates of MN-BI-FGSM integrated with input transformations (i.e., DIM, TIM, SIM, and DTS) are ~12.1%, ~4.9%, ~15.2%, and ~26.3% higher than MN-BI-FGSM on average, respectively. The results indicate that integrating input transformations into MN-BI-FGSM further improves transferability. MN-BI-DTS, which is a combination of MN-BI-FGSM and DTS, shows the best attack success rates. Its white-box attack success rate reaches 100%, and its average black-box attack success rate is ~72.0%.

## 4.3  Attack Ensemble Models

In this experiment, we use Inc-v3, Inc-v4, IncRes-v2, Res-152, Inc-v3$_{adv}$ and IncRes-v2$_{ens}$ as the hold-out model, respectively. The ensemble model is the

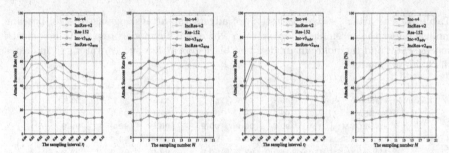

(a)    MN-NI-FGSM (b)    MN-NI-FGSM (c)  ENI-FGSM for (d)  ENI-FGSM for
for various sampling for various sampling various sampling in- various    sampling
intervals          numbers          tervals          numbers

**Fig. 3.** The attack success rates (%) of the adversarial examples generated on Inc-v3 against Inc-v4, IncRes-v2, Res-152, Inc-v3$_{adv}$ and IncRes-v2$_{ens}$ (black-box) for various sampling intervals and various sampling numbers

ensemble of the other five models by ensemble in logits (see Sect. 2.1), and the ensemble weights are all set to $1/5$. We use MN-BI-FGSM to generate adversarial examples on the ensemble models because MN-BI-FGSM shows the best attack success rates on single models. Baselines include FGSM, I-FGSM, and parent attacks of MN-BI-FGSM (i.e., ENI-FGSM and our MN-NI-FGSM). We use our MN-BI-FGSM and baselines to white-box attack the ensemble models and black-box attack the hold-out models. Table 4 shows the success attack rates. The white-box attack success rates of our MN-BI-FGSM are not lower than baselines and not lower than 99.9%. In addition, the black-box attack success rates of our MN-BI-FGSM are $\sim$24.7%, $\sim$35.6%, $\sim$3.0% and $\sim$1.9% higher than one-step attack (i.e., FGSM), basic iterative attack (i.e., I-FGSM) and its parent attacks (i.e., ENI-FGSM and our MN-NI-FGSM) on average.

### 4.4  Ablation Study

We conduct an ablation study here for MN-NI-FGSM. For the sampling distribution, we adopt the linear sampling as in [28].

**On Sampling Interval:** We pre-set $N = 11$ and set $\eta$ from 0 to 0.1 with a step size of 0.01. The results are shown in Fig. 3a. When $\eta \leq 0.02$, the attack success rates show an upward trend in general, and when $\eta \geq 0.02$, the attack success rates show a downward trend in general. Therefore, we set $\eta = 0.02$ for MN-NI-FGSM to achieve the best attack success rate.

**On Sampling Number:** According to the above results, we pre-set $\eta = 0.02$ and set $N$ from 1 to 21 with a step size of 2. The results are shown in Fig. 3b. When $N \leq 11$, the attack success rates show an upward trend in general, and when $N \geq 11$, the attack success rates rise slowly and gradually stabilize. Since the larger $N$, the higher the computational cost, we set $N = 11$ for MN-NI-FGSM.

In summary, we set the sampling interval $\eta = 0.02$ and the sampling number $N = 11$ for MN-NI-FGSM. As shown in Fig. 3c and 3d, we conduct an ablation study for ENI-FGSM, since the values of the sampling interval $\eta$ and the sampling number $N$ of ENI-FGSM are not given in [28]. We can draw the same conclusion as MN-NI-FGSM for ENI-FGSM, i.e., the sampling interval $\eta = 0.02$ and the sampling number $N = 11$. For PN-NI-FGSM, We set the same hyperparameters as EMI-FGSM as in [28] because EMI-FGSM and PN-NI-FGSM all use the previous gradient to describe the neighborhood. Similarly, we set the same hyperparameters for PN-NI-FGSM as ENI-FGSM, and we set the same hyperparameters for BFRVM attacks as their parent attacks.

## 5 Conclusion

In this study, we first propose two methods to improve the transferability of NI-FGSM, i.e., PN-NI-FGSM and MN-NI-FGSM. Both of the above methods show higher black-box attack success rates than NI-FGSM. Then we define the front vector and the rear vector, and we classify existing gradient-based momentum iterative attacks into front vector attacks and rear vector attacks. Since the front vector can refine the pre-update, and the rear vector can modify the pre-update, we propose a method called BFRVM that combines front vector attacks and rear vector attacks. To further improve transferability, we integrate input transformations into the BFRVM attacks and generate adversarial examples on the ensemble models through the BFRVM attacks. The results show that the best white-box attack success rate of our BFRVM attacks reaches 100.0% and the black box attack success rates are higher than the state-of-the-art attacks. Our work also indicates that existing models still need to be more robust to defend against adversarial attacks.

**Acknowledgements.** This work was supported by the National Natural Science Foundation of China (No. 62072250, 62172435, U1804263, U20B2065, 61872203, 71802110 and 61802212), and by the Zhongyuan Science and Technology Innovation Leading Talent Project of China (No. 214200510019), and the Natural Science Foundation of Jiangsu Province (No. BK20200750), and Open Foundation of Henan Key Laboratory of Cyberspace Situation Awareness (No. HNTS2022002), and Post graduate Research & Practice Innvoation Program of Jiangsu Province (No. KYCX200974), and the Opening Project of Guangdong Province Key Laboratory of Information Security Technology (No. 2020B1212060078), and the Ministry of education of Humanities and Social Science project (No. 19YJA630061) and the Priority Academic Program Development of Jiangsu Higher Education Institutions (PAPD) fund.

## References

1. Carlini, N., Wagner, D.: Towards evaluating the robustness of neural networks. In: 2017 IEEE Symposium on Security and Privacy (sp), pp. 39–57. IEEE (2017)
2. Chrysos, G.G., Moschoglou, S., Bouritsas, G., Panagakis, Y., Deng, J., Zafeiriou, S.: P-nets: deep polynomial neural networks. In: Proceedings of the IEEE/CVF Conference on Computer Vision and Pattern Recognition, pp. 7325–7335 (2020)

3. Cococcioni, M., Ruffaldi, E., Saponara, S.: Exploiting posit arithmetic for deep neural networks in autonomous driving applications. In: 2018 International Conference of Electrical and Electronic Technologies for Automotive, pp. 1–6. IEEE (2018)

4. Dong, Y., et al.: Boosting adversarial attacks with momentum. In: Proceedings of the IEEE Conference on Computer Vision and Pattern Recognition, pp. 9185–9193 (2018)

5. Dong, Y., Pang, T., Su, H., Zhu, J.: Evading defenses to transferable adversarial examples by translation-invariant attacks. In: Proceedings of the IEEE/CVF Conference on Computer Vision and Pattern Recognition, pp. 4312–4321 (2019)

6. Franchi, G., et al.: MUAD: multiple uncertainties for autonomous driving benchmark for multiple uncertainty types and tasks. arXiv preprint arXiv:2203.01437 (2022)

7. Ghenescu, V., Mihaescu, R.E., Carata, S.V., Ghenescu, M.T., Barnoviciu, E., Chindea, M.: Face detection and recognition based on general purpose DNN object detector. In: 2018 International Symposium on Electronics and Telecommunications (ISETC), pp. 1–4. IEEE (2018)

8. Goodfellow, I.J., Shlens, J., Szegedy, C.: Explaining and harnessing adversarial examples. arXiv preprint arXiv:1412.6572 (2014)

9. Guo, C., Rana, M., Cisse, M., Van Der Maaten, L.: Countering adversarial images using input transformations. arXiv preprint arXiv:1711.00117 (2017)

10. Hao, C., et al.: NAIS: neural architecture and implementation search and its applications in autonomous driving. In: 2019 IEEE/ACM International Conference on Computer-Aided Design (ICCAD), pp. 1–8. IEEE (2019)

11. He, K., Zhang, X., Ren, S., Sun, J.: Deep residual learning for image recognition. In: Proceedings of the IEEE Conference on Computer Vision and Pattern Recognition, pp. 770–778 (2016)

12. He, K., Zhang, X., Ren, S., Sun, J.: Identity mappings in deep residual networks. In: Leibe, B., Matas, J., Sebe, N., Welling, M. (eds.) ECCV 2016. LNCS, vol. 9908, pp. 630–645. Springer, Cham (2016). https://doi.org/10.1007/978-3-319-46493-0_38

13. Ioffe, S., Szegedy, C.: Batch normalization: accelerating deep network training by reducing internal covariate shift. In: International Conference on Machine Learning, pp. 448–456. PMLR (2015)

14. Krizhevsky, A., Sutskever, I., Hinton, G.E.: ImageNet classification with deep convolutional neural networks. Adv. Neural Inf. Process. Syst. **25** (2012)

15. Kurakin, A., et al.: Adversarial attacks and defences competition. In: Escalera, S., Weimer, M. (eds.) The NIPS '17 Competition: Building Intelligent Systems. TSSCML, pp. 195–231. Springer, Cham (2018). https://doi.org/10.1007/978-3-319-94042-7_11

16. Kurakin, A., Goodfellow, I., Bengio, S., et al.: Adversarial examples in the physical world (2016)

17. Liao, F., Liang, M., Dong, Y., Pang, T., Hu, X., Zhu, J.: Defense against adversarial attacks using high-level representation guided denoiser. In: Proceedings of the IEEE Conference on Computer Vision and Pattern Recognition, pp. 1778–1787 (2018)

18. Lin, J., Song, C., He, K., Wang, L., Hopcroft, J.E.: Nesterov accelerated gradient and scale invariance for adversarial attacks. arXiv preprint arXiv:1908.06281 (2019)

19. Madry, A., Makelov, A., Schmidt, L., Tsipras, D., Vladu, A.: Towards deep learning models resistant to adversarial attacks. arXiv preprint arXiv:1706.06083 (2017)

20. Nesterov, Y.: A method for unconstrained convex minimization problem with the rate of convergence o $\left(\frac{1}{k^2}\right)$ (1983)

21. Polyak, B.T.: Some methods of speeding up the convergence of iteration methods. USSR Comput. Math. Math. Phys. **4**(5), 1–17 (1964)
22. Russakovsky, O., et al.: ImageNet large scale visual recognition challenge. Int. J. Comput. Vision **115**(3), 211–252 (2015)
23. Szegedy, C., Ioffe, S., Vanhoucke, V., Alemi, A.A.: Inception-v4, inception-resnet and the impact of residual connections on learning. In: Thirty-First AAAI Conference on Artificial Intelligence (2017)
24. Szegedy, C., et al.: Going deeper with convolutions. In: Proceedings of the IEEE Conference on Computer Vision and Pattern Recognition, pp. 1–9 (2015)
25. Szegedy, C., Vanhoucke, V., Ioffe, S., Shlens, J., Wojna, Z.: Rethinking the inception architecture for computer vision. In: Proceedings of the IEEE Conference on Computer Vision and Pattern Recognition, pp. 2818–2826 (2016)
26. Szegedy, C., et al.: Intriguing properties of neural networks. arXiv preprint arXiv:1312.6199 (2013)
27. Tramèr, F., Kurakin, A., Papernot, N., Goodfellow, I., Boneh, D., McDaniel, P.: Ensemble adversarial training: attacks and defenses. arXiv preprint arXiv:1705.07204 (2017)
28. Wang, X., Lin, J., Hu, H., Wang, J., He, K.: Boosting adversarial transferability through enhanced momentum. arXiv preprint arXiv:2103.10609 (2021)
29. Xie, C., Wang, J., Zhang, Z., Ren, Z., Yuille, A.: Mitigating adversarial effects through randomization. arXiv preprint arXiv:1711.01991 (2017)
30. Xie, C., et al.: Improving transferability of adversarial examples with input diversity. In: Proceedings of the IEEE/CVF Conference on Computer Vision and Pattern Recognition, pp. 2730–2739 (2019)
31. Zhang, J., Wang, J., Wang, H., Luo, X.: Self-recoverable adversarial examples: a new effective protection mechanism in social networks. IEEE Trans. Circuits Syst. Video Technol. 1 (2022). https://doi.org/10.1109/TCSVT.2022.3207008

# Manipulated Face Detection and Localization Based on Semantic Segmentation

Gen Li[1,2], Xianfeng Zhao[1,2(✉)], Yun Cao[1,2], and Chengqiao Hu[1,2]

[1] State Key Laboratory of Information Security, Institute of Information Engineering, Chinese Academy of Sciences, Beijing 100195, China
{ligen1,zhaoxianfeng,caoyun,huchengqiao}@iie.ac.cn
[2] School of Cyber Security, University of Chinese Academy of Sciences, Beijing 100195, China

**Abstract.** In this paper, we propose a novel manipulated face detection and localization approach, which simultaneously detect manipulated face images and videos and locate the manipulated regions at semantic-level. To do this, we design a multi-branch autoencoder composed of four types of modules, including feature encoder, shared decoder, semantic decoder, and classification network. The feature encoder extracts latent feature from the input face image. The shared decoder obtains structure feature from the latent feature. The four semantic decoders decode structure feature into four different semantic prediction masks, respectively. The classification network outputs the semantic prediction labels based on the latent feature. Finally, the manipulation prediction label and manipulation prediction mask of the input face image can be generated with the semantic prediction labels and semantic prediction masks. Extensive experiments show that our approach can effectively detect and locate manipulated face images and videos at semantic-level, even under cross-manipulation, cross-dataset, and cross-compression scenarios.

**Keywords:** Manipulated face detection · Manipulated face localization · Semantic segmentation

## 1 Introduction

The proliferation of artificial intelligence has given rise to various face image and video tampering methods, especially generative models [1,2]. Since these techniques can synthesize more and more realistic face images and videos that are hardly distinguishable by humans [3,4]. Abuse of manipulated face images and videos can easily trigger severe societal problems or political threats over the world. Therefore, it is important to develop effective methods for exposing face manipulation.

The two major concerns in the field of manipulated face forensics are manipulated face detection and manipulated face localization. Manipulated face detection methods can classify face images and videos into fake face images and

X. Zhao et al. (Eds.): IWDW 2022, LNCS 13825, pp. 98–113, 2023.
https://doi.org/10.1007/978-3-031-25115-3_7

videos or real face images and videos. As the release of large-scale face manipulation datasets [5–7], it enables the training of Deep Neural Networks (DNN) to identify manipulated face images and videos with various deep features [8–13]. Manipulated face localization methods with the goal to locate the manipulated regions. Early works focus on three commonly used means of tampering, including removal, copy-move, and splicing [14–16]. They are not well suited for locating advanced face manipulations. So far, only several works have been proposed in the field of manipulated face localization [17–20]. However, all the existing manipulated face localization methods can only locate manipulated regions at pixel-level. None of them considers locating manipulated regions at semantic-level, which is more significant and valuable in the field of manipulated face forensics.

In this paper, we propose a novel approach for simultaneously performing detection and localization of manipulated face images and videos at semantic-level. Our designed multi-branch autoencoder comprises a feature encoder, a shared decoder, four semantic decoders, and a classification network. For an input face image, firstly, we encode it into the latent feature by the feature encoder. Second, we decode the latent feature into structure feature by the shared decoder. Third, we feed the structure feature into four semantic decoders for generating four semantic prediction masks, respectively. Meanwhile, we feed the latent feature into the classification network for obtaining the semantic prediction labels. Lastly, the manipulation prediction label and manipulation prediction mask of the input face image can be generated with the semantic prediction labels and semantic prediction masks.

To the best of our knowledge, this is the first solution to deal with the problem of manipulated face detection and localization at semantic-level. We experimentally evaluate the effectiveness of our proposed approach on five widely used face manipulation datasets. The experimental results show that our approach can accurately detect and locate manipulated face images and videos at semantic-level. Moreover, the performance of inter-dataset evaluation demonstrates that our approach is effective even under cross-manipulation, cross-dataset, and cross-compression scenarios.

The rest of the paper is organized as follows. Section 2 provides an overview about our work. Section 3 describes the overall pipeline of our proposed approach and the details of all parts. The experimental results are presented in Sect. 4. Lastly, Sect. 5 concludes our work.

## 2 Related Works

### 2.1 Face Manipulation

Face Manipulation has been studied for a long period. The methods can be mainly divided into four types regarding the level of manipulation. They are entire face synthesis, attribution manipulation, expression swap, and identity swap. Entire face synthesis creates entire non-existent face images, usually through powerful Generative Adversarial Networks (GAN), such as ProGAN,

StyleGAN, and CycleGAN [21–23]. Attribution manipulation, also known as face editing or face retouching, consists of modifying some attributes of the face such as the color of the hair, the gender, the age, and adding glasses. This manipulation process is usually carried out through IcGAN, StarGAN, AttGAN, and STGAN [24–26]. Expression swap, also known as face reenactment, consists of modifying the facial expression of the person. Face2Face and NeuralTextures are two representative techniques [3,4]. Identity swap is the most popular face manipulation type, which consists of replacing the face of one person in an image or a video with the face of another person. Classical computer graphics technique and deep learning based techniques known DeepFake are usually considered in this field [27,28]. Based on the different face manipulation types, the manipulated face images and videos contain fake information in the different facial regions.

## 2.2 Manipulated Face Detection

Manipulated face detection is normally deemed a binary classification problem. In general, the methods can be divided into three categories based on the partition of feature space. They are spatial feature based methods, temporal feature based methods, and multi-domain feature based methods. Spatial feature based methods exploit the artifacts, such as visual artifacts, inconsistent head pose, missing symmetry, warping artifacts, and convolutional traces, to expose manipulated face images [13,29–32]. Various types of DNN, such as capsule network, residual network, and attention mechanism, were also applied to automatically learn salient and discriminative features to detect manipulated face images [11,33]. Temporal feature based methods exploit the anomalies between consecutive frames, such as optical flow motion, eye blinking pattern, facial action units, and prediction error, to expose manipulated face videos [12,34]. Multi-domain feature based methods exploit multi-stream neural networks to combine spatial feature, temporal feature, and frequency feature for exposing manipulated face videos [35]. The impressive progress has been made in the performance of detection of manipulated face images and videos under the proposed countermeasures.

## 2.3 Manipulated Face Localization

Manipulated face localization aims to locate the manipulated regions of face images and videos. A multi-task learning approach is designed for simultaneously detecting manipulated face images and videos and locating the manipulated regions [17]. The network includes an encoder and a Y-shaped decoder. The encoder is used for the binary classification. One branch of the decoder is used for segmenting the manipulated regions while the other branch is used for reconstructing the input. Some researchers introduced the attention mechanism into encoder-decoder for capturing the fake textures of manipulated face images [1,19]. Another method called face X-ray only assumes the existence of a blending step in the face image synthesis process [18]. The face X-ray shows the blending boundary for a manipulated face image and the absence of blending

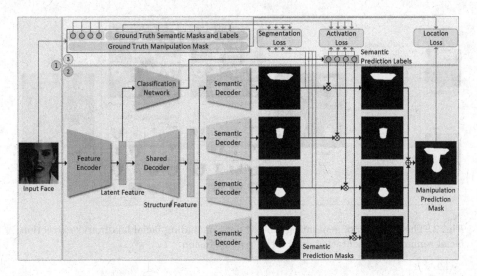

**Fig. 1.** The overall pipeline of our proposed approach.

for a real face image. Besides, adding the information of facial landmarks into the input face images can also improve the performance of manipulation mask prediction [20]. All the existing manipulated face localization methods focus on locating the manipulated regions at pixel-level. However, none of them considers locating the manipulated regions at semantic-level which is the focus of our work.

## 3 Methodology

In this section, we describe our proposed approach from three main parts in detail, including labeled data generation, multi-branch autoencoder, and loss function. Figure 1 graphically shows the overall pipeline of our approach.

### 3.1 Labeled Data Generation

In order to generate labeled data, including face images, semantic masks, semantic labels, and manipulation masks, we carry out three data processing steps of facial landmarks extraction, facial semantic segmentation, and facial regions fusion. Figure 2 shows the overview of generating labeled data.

**Facial Landmarks Extraction.** Facial landmarks are locations on the face carrying important structural information of facial features. In our method, we extract 68 facial landmarks, from number 0 to 67, for each face image through the toolbox *dlib*. An example of facial landmarks is shown in Fig. 2(a). For each face image, the facial region is cropped and resized to the uniform size with 256 × 256 pixels in RGB channels.

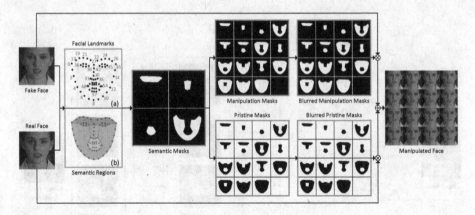

**Fig. 2.** The overview of generating labeled data, including facial landmarks extraction, facial semantic segmentation, and facial regions fusion.

**Facial Semantic Segmentation.** According to the location of facial features, in our method, we segment each face image into four different semantic regions, including eyes, nose, mouth, and rest. An example of semantic regions is shown in Fig. 2(b). The specific facial landmarks considered in the segmentation process of semantic regions are

- *Eyes:* facial landmarks from number 17 to 26, and number 36 to 47.
- *Nose:* facial landmarks from number 27 to 35, and number 39, 42.
- *Mouth:* facial landmarks from number 31 to 35, and number 48 to 67.
- *Rest:* facial landmarks from number 0 to 16, and number 17 to 26.

Then, the semantic masks of eyes $M_e$, nose $M_n$, mouth $M_m$, and rest $M_r$ are defined as

$$\begin{cases} M_e = f_{CH}(J_e) \\ M_n = f_{CH}(J_n) \\ M_m = f_{CH}(J_m) \\ M_r = f_{CH}(J_r) - (M_e + M_n + M_m) \end{cases}, \tag{1}$$

where each pixel value in the semantic masks has binary value of 0 or 1. $f_{CH}$ denotes the operation of Convex Hull. $J_e$, $J_n$, $J_m$, and $J_r$ are the sets of facial landmarks of eyes, nose, mouth, and rest. Therefore, the semantic regions of eyes $R_e$, nose $R_n$, mouth $R_m$, and rest $R_r$ can be obtained through

$$\begin{cases} R_e = I \odot M_e \\ R_n = I \odot M_n \\ R_m = I \odot M_m \\ R_r = I \odot M_r \end{cases}, \tag{2}$$

where $I$ is the input face image. $\odot$ denotes the operation of element-wise multiplication.

**Facial Regions Fusion.** In order to generate manipulated face images with different manipulation regions, we obtain manipulation masks through combining four semantic masks with each other. The manipulation mask $M_M$ is defined as

$$M_M = l_e \cdot M_e + l_n \cdot M_n + l_m \cdot M_m + l_r \cdot M_r, \tag{3}$$

where $l_e$, $l_n$, $l_m$, and $l_r$ are the semantic labels of eyes, nose, mouth, and rest. They have binary value of 0 or 1. When the value of semantic label is equal to 1, the corresponding semantic mask is activated and vice versa.

Then, the manipulated face image $I_M$ can be generated through

$$I_M = I_F \odot f_{GB}(M_M) + I_R \odot f_{GB}(1 - M_M), \tag{4}$$

where $I_F$ and $I_R$ are paired fake face image and real face image. It should be noted that the paired fake face image and real face image share the same background and facial landmarks. These data can be collected from the public face manipulation datasets. $f_{GB}$ denotes the operation of Gaussion Blur for turning a binary mask into a soft mask, which is helpful to weaken the artifacts of image fusion boundary. According to the value of semantic labels, the manipulation label $l_M$ of the manipulated face image $I_M$ is defined as

$$l_M = f_{MAX}(l_e, l_n, l_m, l_r), \tag{5}$$

where $f_{MAX}$ denotes the operation of Maximizing. Exceptionally, the manipulation mask $M_M$ is a trivial blank image with all pixel values equal to 0 when all semantic labels are set to 0. In this case, the manipulated face image $I_M$ will be transformed into the pristine face image.

As described above, we are able to produce a huge number of labeled data step by step. In practice, we generate the labeled data dynamically along with the training process.

## 3.2   Multi-branch Autoencoder

In order to detect and locate manipulated face images and videos at semantic-level, we designed a multi-branch autoencoder to predict manipulation probability and locate manipulation mask of the input face image. The multi-branch autoencoder consists of four types of modules, feature encoder, shared decoder, semantic decoder, and classification network.

**Feature Encoder.** The function of feature encoder $E_F$ is encoding the input face image $I$ into the latent feature $F_l$, which can be formally described as

$$F_l = E_F(I), \tag{6}$$

where $F_l$ contains the category information and structure information of the input face image. The sizes of $I$ and $F_l$ are $256 \times 256 \times 3$ and $16 \times 16 \times 128$. $E_F$ consists of nine convolutional (Conv) layers. Following each Conv layer is a batch normalization (BatchNorm) layer and a rectified linear unit (ReLU).

**Shared Decoder.** The function of shared decoder $D_{SH}$ is decoding the latent feature $F_l$ into the structure feature $F_s$, which can be formally described as

$$F_s = D_{SH}(F_l),  \tag{7}$$

where $F_s$ contains the semantic information of the input face image. The sizes of $F_s$ is $64 \times 64 \times 32$. $D_{SH}$ consists of four trans convolutional (TransConv) layers. Following each TransConv layer is a BatchNorm layer and a ReLU.

**Semantic Decoder.** In order to generate semantic prediction masks of eyes $\hat{M}_e$, nose $\hat{M}_n$, mouth $\hat{M}_m$, and rest $\hat{M}_r$, we constructed four identical semantic decoders $D_{SEe}$, $D_{SEn}$, $D_{SEm}$, and $D_{SEr}$ for decoding the structure feature $F_s$ respectively. The generation process of the semantic prediction masks can be formally described as

$$\begin{cases} F_e = [F_{e0}, F_{e1}] = D_{SEe}(F_s) \\ F_n = [F_{n0}, F_{n1}] = D_{SEn}(F_s) \\ F_m = [F_{m0}, F_{m1}] = D_{SEm}(F_s) \\ F_r = [F_{r0}, F_{r1}] = D_{SEr}(F_s) \end{cases}  \tag{8}$$

$$\begin{cases} \hat{M}_e = f_{RO}(F_{e1}) \\ \hat{M}_n = f_{RO}(F_{n1}) \\ \hat{M}_m = f_{RO}(F_{m1}) \\ \hat{M}_r = f_{RO}(F_{r1}) \end{cases}  \tag{9}$$

where $F_e$, $F_n$, $F_m$, and $F_r$ are activation features of $\hat{M}_e$, $\hat{M}_n$, $\hat{M}_m$, and $\hat{M}_r$. They have the same size of $256 \times 256 \times 2$. $f_{RO}$ denotes the operation of Rounding Off. $\hat{M}_e$, $\hat{M}_n$, $\hat{M}_m$, and $\hat{M}_r$ have the same size of $256 \times 256 \times 1$. All pixel values in the semantic prediction masks are either 0 or 1. Each semantic decoder consists of five TransConv layers and end up with a softmax activation function. The first four TransConv layers are followed by a BatchNorm layer and a ReLU.

**Classification Network.** In order to obtain semantic prediction labels of eyes $\hat{l}_e$, nose $\hat{l}_n$, mouth $\hat{l}_m$, and rest $\hat{l}_r$, we constructed a classification network $N_C$ for extracting category information of semantic regions from the latent feature $F_l$. The classification process of semantic regions can be formally described as

$$[p_e, p_n, p_m, p_r] = N_C(F_l),  \tag{10}$$

$$\left[\hat{l}_e, \hat{l}_n, \hat{l}_m, \hat{l}_r\right] = f_{RO}([p_e, p_n, p_m, p_r]),  \tag{11}$$

where $p_e$, $p_n$, $p_m$, and $p_r$ are manipulation probabilities of eyes, nose, mouth, and rest. They are within the range of $[0, 1]$. $\hat{l}_e$, $\hat{l}_n$, $\hat{l}_m$, and $\hat{l}_r$ have binary value of 0 or 1. When the value of semantic prediction label is equal to 1, the corresponding semantic region is classified as manipulation region and vice versa. $N_C$ consists of six Conv layers and end up with a sigmoid activation function. The first five Conv layers are followed by a BatchNorm layer and a ReLU.

According to the semantic prediction labels and semantic prediction masks, the manipulation prediction label $\hat{l}_M$ and manipulation prediction mask $\hat{M}_M$ of the input image $I$ can be obtained by

$$\hat{l}_M = f_{MAX}\left(\hat{l}_e, \hat{l}_n, \hat{l}_m, \hat{l}_r\right), \tag{12}$$

$$\hat{M}_M = \hat{l}_e \cdot \hat{M}_e + \hat{l}_n \cdot \hat{M}_n + \hat{l}_m \cdot \hat{M}_m + \hat{l}_r \cdot \hat{M}_r. \tag{13}$$

Therefore, for each input face image, we can predict the manipulation probability and locate the manipulated regions at semantic-level through our designed multi-branch autoencoder.

From the view of the overall structure of multi-branch autoencoder, the information extracted from the modules of feature encoder and shared decoder are shared between the tasks of manipulated face detection and localization, which is helpful to improve the overall performance of the network.

### 3.3   Loss Function

During the training process of multi-branch autoencoder, we adopt three types of loss function to optimize the model. They are segmentation loss, activation loss and location loss.

**Segmentation Loss.** The segmentation loss $L_{seg}$ is used to measure the agreement between the semantic prediction mask and the ground truth semantic mask, which is defined as

$$L_{seg} = -\frac{1}{N}\sum_{k=1}^{N}\frac{1}{H\times W}\sum_{(i,j)=(0,0)}^{(H,W)}\ell_{(i,j)}^k, \tag{14}$$

where $N$ is the number of samples. $k$ is the sample index. $H \times W$ is the size of semantic mask. $(i,j)$ is the index denoting the metadata location. $\ell_{(i,j)}^k$ denotes the agreement between the pixel $b_{(i,j)}^k$ in the ground truth semantic mask and the element $\hat{b}_{(i,j)}^k$ in the activation feature of semantic prediction mask, which is defined as

$$\ell_{(i,j)}^k = b_{(i,j)}^k \log \hat{b}_{(i,j)}^k + \left(1 - b_{(i,j)}^k\right) \log \left(1 - \hat{b}_{(i,j)}^k\right), \tag{15}$$

where $b_{(i,j)}^k$ and $\hat{b}_{(i,j)}^k$ can be obtained through Eq. (1) and Eq. (8).

**Activation Loss.** The activation loss $L_{act}$ is used to measure the bias between the manipulation probability $p^k$ and the ground truth semantic label $l^k$, which is defined as

$$L_{act} = \frac{1}{N}\sum_{k=1}^{N}|l^k - p^k|, \tag{16}$$

where $p^k$ can be obtained through Eq. (10).

**Location Loss.** The location loss $L_{loc}$ is used to measure the agreement between the manipulation prediction mask and the ground truth manipulation mask, which is defined as

$$L_{loc} = \frac{1}{N} \sum_{k=1}^{N} \frac{1}{H \times W} \sum_{(i,j)=(0,0)}^{(H,W)} \left| c_{(i,j)}^{k} - \hat{c}_{(i,j)}^{k} \right|, \qquad (17)$$

where $c_{(i,j)}^{k}$ and $\hat{c}_{(i,j)}^{k}$ are pixel values in the ground truth manipulation mask and manipulation prediction mask, which can be obtained through Eq. (3) and Eq. (13).

The total loss is the weighted sum of the three losses, which is defined as

$$L = \frac{\alpha}{4} \sum_{v \in \{e,n,m,r\}} L_{seg}^{v} + \frac{\beta}{4} \sum_{v \in \{e,n,m,r\}} L_{act}^{v} + \gamma L_{loc}, \qquad (18)$$

where $v$ denotes the semantic index. $\alpha$, $\beta$, and $\gamma$ are the weights balancing the three losses. In our method, we set the three weights equal to 1 for the reason that the classification task and the localization task are equally important.

## 4   Experiments

### 4.1   Experimental Setup

**Datasets.** In our experiments, we evaluate our method on five widely used face manipulation datasets, including FaceForensics++ [5], Celeb-DF [6], DFD [5], DFDC [7], and UADFV [29]. FaceForensics++ contains 1,000 real videos and each one corresponds to four fake videos. The fake videos are generated using four different face manipulation methods, including DeepFake (FF++DF), Face2Face (FF++F2F), FaceSwap (FF++FS), and NeuralTextures (FF++NT). Each subdataset contains three video compression quality, including high quality with Constant Rate Factor (CRF) equal to 0, middle quality with CRF equal to 23, and low quality with CRF equal to 40. Celeb-DF contains 590 real videos and 5,639 fake videos. DFD contains 363 real videos and 3,068 fake videos. DFDC contains more than 10,000 real videos and 100,000 fake videos. UADFV contains 49 real videos and 49 fake videos.

We randomly sample equal number of pristine videos and manipulated video from each dataset. Three fourth of videos are taken as training data and the rest of videos as testing data.

**Implementation Details.** In the training process, the batch size is set to 150 and the total number of epochs is set to 100. The learning rate is set as $10^{-3}$ using Adam optimizer at first and then is linearly decayed to $10^{-6}$ every 25 epochs. All training and testing experiments are conducted on one NVIDIA Tesla-P100 12G GPU.

**Table 1.** Manipulated face localization error (%) at pixel-level and semantic-level.

| $l_e l_n l_m l_r$ | 1000 | 0100 | 0010 | 0001 | 1100 | 1010 | 1001 | 0110 | 0101 | 0011 | 1110 | 1101 | 1011 | 0111 | 1111 | 0000 |
|---|---|---|---|---|---|---|---|---|---|---|---|---|---|---|---|---|
| Dataset | $ERR_{pix}$ | | | | | | | | | | | | | | | |
| FF++DF | 00.71 | 00.38 | 00.42 | 02.32 | 00.94 | 01.08 | 02.54 | 00.65 | 02.34 | 02.15 | 01.22 | 02.43 | 02.36 | 02.07 | 02.19 | 00.06 |
| FF++F2F | 01.21 | 00.92 | 01.02 | 03.04 | 01.50 | 01.77 | 03.38 | 01.36 | 03.33 | 03.00 | 01.97 | 03.49 | 03.34 | 03.16 | 03.29 | 00.56 |
| FF++FS | 00.89 | 00.67 | 00.74 | 02.47 | 01.20 | 01.32 | 02.61 | 00.98 | 02.56 | 02.27 | 01.52 | 02.55 | 02.42 | 02.20 | 02.21 | 00.35 |
| FF++NT | 00.86 | 00.56 | 00.62 | 02.48 | 01.10 | 01.32 | 02.72 | 00.88 | 02.58 | 02.33 | 01.44 | 02.64 | 02.59 | 02.31 | 02.41 | 00.25 |
| Celeb-DF | 06.42 | 06.78 | 06.85 | 06.59 | 05.76 | 05.31 | 06.31 | 05.82 | 06.36 | 04.94 | 05.05 | 06.63 | 04.82 | 05.14 | 05.48 | 08.62 |
| DFD | 03.62 | 03.68 | 03.56 | 06.30 | 03.92 | 03.30 | 06.70 | 02.97 | 05.57 | 05.58 | 03.80 | 06.52 | 06.12 | 04.97 | 06.03 | 06.91 |
| DFDC | 08.02 | 07.32 | 07.86 | 11.60 | 08.40 | 07.90 | 10.87 | 07.02 | 10.37 | 11.21 | 08.31 | 11.16 | 10.59 | 10.35 | 11.58 | 08.80 |
| UADFV | 01.48 | 01.28 | 01.12 | 02.91 | 02.09 | 02.15 | 03.20 | 01.97 | 03.17 | 02.78 | 02.84 | 03.33 | 03.06 | 03.32 | 03.25 | 00.47 |
| Dataset | $ERR_{sem}$ | | | | | | | | | | | | | | | |
| FF++DF | 00.39 | 00.34 | 00.36 | 00.38 | 00.66 | 00.59 | 00.61 | 00.60 | 00.58 | 00.62 | 00.89 | 00.87 | 00.84 | 00.86 | 01.19 | 00.18 |
| FF++F2F | 01.83 | 02.19 | 02.34 | 01.70 | 02.73 | 02.81 | 02.33 | 03.15 | 03.00 | 02.97 | 03.82 | 03.67 | 03.54 | 04.27 | 05.03 | 01.69 |
| FF++FS | 00.84 | 00.96 | 01.24 | 01.02 | 01.26 | 01.13 | 01.16 | 01.24 | 01.28 | 01.30 | 01.59 | 01.55 | 01.42 | 01.60 | 01.93 | 01.01 |
| FF++NT | 00.82 | 00.88 | 00.90 | 00.72 | 01.21 | 01.21 | 01.07 | 01.29 | 01.15 | 01.14 | 01.74 | 01.50 | 01.52 | 01.72 | 02.19 | 00.73 |
| Celeb-DF | 09.38 | 13.30 | 08.75 | 09.78 | 11.22 | 04.69 | 06.77 | 09.56 | 10.14 | 04.44 | 08.49 | 09.90 | 02.61 | 07.21 | 08.23 | 14.92 |
| DFD | 05.48 | 04.54 | 05.39 | 04.94 | 06.44 | 05.27 | 05.80 | 03.98 | 04.20 | 04.94 | 06.43 | 07.73 | 06.24 | 04.94 | 09.36 | 09.77 |
| DFDC | 14.04 | 13.55 | 15.89 | 17.20 | 17.74 | 14.07 | 14.81 | 14.64 | 13.91 | 16.61 | 19.29 | 17.96 | 15.06 | 15.52 | 20.93 | 18.51 |
| UADFV | 02.83 | 04.71 | 03.18 | 00.70 | 06.76 | 04.75 | 01.15 | 07.37 | 02.71 | 01.74 | 09.65 | 03.81 | 01.93 | 05.57 | 05.83 | 01.21 |

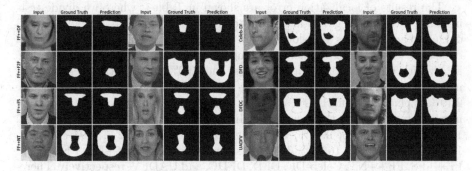

**Fig. 3.** The examples of manipulated face localization.

**Evaluation Metrics.** We report the experimental results of manipulated face detection and localization with the metrics of accuracy $ACC$, pixel-level error $ERR_{pix}$ and semantic-level error $ERR_{sem}$. For each input face image, the two localization errors can be calculated by

$$ERR_{pix} = \frac{1}{H \times W} \sum_{(i,j)=(0,0)}^{(H,W)} \left| c_{(i,j)} - \hat{c}_{(i,j)} \right|, \quad (19)$$

$$ERR_{sem} = \frac{1}{4} \sum_{v \in \{e,n,m,r\}} \left| l_v - \hat{l}_v \right|, \quad (20)$$

where the symbols of $H$, $W$, $(i,j)$, $c_{(i,j)}$, $\hat{c}_{(i,j)}$, $v$, $l_v$ and $\hat{l}_v$ have been mentioned in Eq. (3), Eq. (11), Eq. (14), Eq. (17) and Eq. (20).

**Table 2.** Manipulated face detection accuracy (%) at semantic-level.

| Dataset | $ACC_e$ | $ACC_n$ | $ACC_m$ | $ACC_r$ | $ACC_M$ |
|---------|---------|---------|---------|---------|---------|
| FF++DF | 99.37 | 99.35 | 99.36 | 99.39 | 99.22 |
| FF++F2F | 97.21 | 96.67 | 95.88 | 98.43 | 95.75 |
| FF++FS | 98.93 | 98.27 | 98.68 | 98.96 | 97.71 |
| FF++NT | 98.54 | 98.70 | 98.53 | 99.24 | 98.08 |
| Celeb-DF | 85.65 | 79.91 | 80.21 | 82.33 | 77.36 |
| DFD | 93.56 | 90.79 | 94.76 | 88.98 | 82.29 |
| DFDC | 79.45 | 77.13 | 73.13 | 77.31 | 70.30 |
| UADFV | 96.92 | 91.63 | 95.99 | 99.52 | 95.91 |

## 4.2 Intra-dataset Evaluation

In the experiments of intra-dataset evaluation, our model is trained on the training data, and performance is evaluated with the corresponding testing data.

The results of localization error are shown in Table 1, where we divide testing data into 16 categories according to the semantic labels. As can be seen, our approach achieves well localization performance on all face manipulation datasets used in our experiments. Taking FF++DF dataset as an example, the localization errors are 0.65% at pixel-level and 0.60% at semantic-level when the manipulation semantic regions are nose and mouth. The localization errors are 2.19% at pixel-level and 1.19% at semantic-level when the whole facial region are manipulated. Overall, the average localization errors are 1.49%, 2.27%, 1.68%, 1.69%, 6.06%, 4.97%, 9.46%, 2.40% at pixel-level and 0.62%, 2.94%, 1.28%, 1.24%, 8.71%, 5.96%, 16.23%, 3.99% at semantic-level on FF++DF, FF++F2F, FF++FS, FF++NT, Celeb-DF, DFD, DFDC, and UADFV, respectively. Therefore, our approach can precisely locate the manipulated facial regions at pixel-level and semantic-level. Figure 3 shows the examples of manipulated face localization.

The results of detection accuracy are shown in Table 2, where $ACC_e$, $ACC_n$, $ACC_m$, $ACC_r$, and $ACC_M$ denote the accuracy of semantic prediction labels $\hat{l}_e$, $\hat{l}_n$, $\hat{l}_m$, $\hat{l}_r$ and manipulation prediction label $\hat{l}_M$. Taking UADFV dataset as an example, the detection accuracy is 96.92% when the manipulation semantic regions contain eyes. The detection accuracy is 95.91% when the face images contain at last one manipulation semantic region. Therefore, our approach can differentiate pristine face images and manipulated face images effectively. Note that the results reported for Celeb-DF dataset and DFDC dataset are comparably lower, due to the fact that these datasets contain more realistic face images generated through the advanced face manipulation methods.

In order to compare the performance of our approach with the existing manipulated face localization methods, we present the experimental results of two other representative methods based on FF++FS dataset in Table 3. We can see that, our approach reduces the localization error from 1.54% to 1.32% at pixel-level when the manipulation semantic regions are eyes and mouth. Overall, our approach reduces the average localization error from 1.99% to 1.68% at pixel-level on all testing data. Thus, our approach outperforms the state-of-the-art manipulated face localization methods. All the existing manipulated face localization methods can locate the manipulated regions only at pixel-level and not at semantic-level.

### 4.3  Inter-dataset Evaluation

In the experiments of inter-dataset evaluation, we perform cross-manipulation, cross-dataset, and cross-compression tests in verification of model generalization.

**Cross-Manipulation Evaluation.** In order to evaluate the generalization ability of our approach under cross-manipulation scenario, we train the model on FF++F2F dataset and test it on three other datasets, including FF++DF, FF++FS, and FF++NT, generated using three different face manipulation

**Table 3.** Manipulated face localization error (%) comparison with the existing methods at pixel-level based on FF++FS dataset.

| $l_e l_n l_m l_r$ | 1000 | 0100 | 0010 | 0001 | 1100 | 1010 | 1001 | 0110 | 0101 | 0011 | 1110 | 1101 | 1011 | 0111 | 1111 | 0000 |
|---|---|---|---|---|---|---|---|---|---|---|---|---|---|---|---|---|
| Method | $ERR_{pix}$ | | | | | | | | | | | | | | | |
| Y.S. [17] | 01.68 | 01.24 | 01.40 | 03.38 | 01.97 | 02.13 | 03.34 | 01.71 | 03.33 | 03.15 | 02.45 | 03.35 | 03.23 | 03.09 | 03.71 | 00.43 |
| F.L. [19] | 01.14 | 00.86 | 00.90 | 02.78 | 01.39 | 01.54 | 02.98 | 01.12 | 02.90 | 02.71 | 01.65 | 02.99 | 02.91 | 02.73 | 02.84 | 00.40 |
| **Ours** | 00.89 | 00.67 | 00.74 | 02.47 | 01.20 | 01.32 | 02.61 | 00.98 | 02.56 | 02.27 | 01.52 | 02.55 | 02.42 | 02.20 | 02.21 | 00.35 |

**Table 4.** Performance (%) of inter-dataset evaluation.

| Training set | Testing set | $ERR_{pix}$ | $ERR_{sem}$ | $ACC_M$ |
|---|---|---|---|---|
| FF++F2F | FF++DF | 05.03 | 05.91 | 95.53 |
| | FF++FS | 14.61 | 27.86 | 73.70 |
| | FF++NT | 04.87 | 07.75 | 93.16 |
| Celeb-DF | DFD | 11.83 | 21.56 | 71.25 |
| | DFDC | 13.40 | 24.20 | 65.34 |
| | UADFV | 09.41 | 15.92 | 81.55 |
| FF++FS-HQ | FF++FS-HQ | 01.68 | 01.28 | 98.65 |
| | FF++FS-MQ | 10.74 | 17.29 | 83.12 |
| | FF++FS-LQ | 14.81 | 24.02 | 72.18 |
| FF++FS-MQ | FF++FS-HQ | 06.37 | 09.20 | 95.45 |
| | FF++FS-MQ | 07.11 | 10.70 | 95.03 |
| | FF++FS-LQ | 10.43 | 17.40 | 91.38 |
| FF++FS-LQ | FF++FS-HQ | 06.77 | 11.79 | 95.63 |
| | FF++FS-MQ | 06.53 | 11.35 | 95.49 |
| | FF++FS-LQ | 04.00 | 06.56 | 97.22 |

methods. The experimental results are shown in Table 4. Compared with the intra-dataset evaluation, the average localization errors on FF++DF dataset and FF++NT dataset are 5.91% and 7.75% at semantic-level, which only rise by 5.29% and 6.51% respectively. The detection accuracy on FF++DF dataset and FF++NT dataset are 95.53% and 93.16%, which only fall by 3.69% and 4.92% respectively. The experimental results demonstrate that our approach can effectively generalize across different face manipulation methods.

**Cross-Dataset Evaluation.** In order to evaluate the generalization ability of our approach under cross-dataset scenario, we train the model on Celeb-DF dataset and test it on three other face swapping datasets, including DFD, DFDC, and UADFV. The experimental results are shown in Table 4. Compared with the intra-dataset evaluation, the average localization errors on DFD dataset and UADFV dataset are 21.56% and 15.92% at semantic-level, which rise by 15.60% and 11.93% respectively, The detection accuracy on DFD dataset and UADFV dataset are 71.25% and 81.55%, which fall by 11.04% and 14.36% respectively. The performance degradation under cross-dataset scenario is likely due to the large difference between training data and testing data.

**Cross-Compression Evaluation.** In order to evaluate the generalization ability of our approach under cross-compression scenario, we train three models on FF++FS high quality (FF++FS-HQ) dataset, FF++FS middle quality (FF++FS-MQ) dataset, and FF++FS low quality (FF++FS-LQ) dataset, respectively. Then, we test them on all three datasets with different video compression quality. The experimental results are shown in Table 4. Taking the model trained on FF++FS-MQ dataset as an example, compared with the intra-dataset evaluation, the average localization error on FF++FS-LQ dataset is 17.40% at semantic-level, which only rise by 6.7%. The detection accuracy on FF++FS-LQ dataset is 91.38%, which only fall by 3.65%. From the results of the model trained on FF++FS-LQ dataset, we can see that our approach can also extract effective feature to detect and locate manipulated face images based on the low quality videos. In general, our approach is robust to the operation of video compression.

## 5    Conclusion

In this paper, we proposed a novel approach for simultaneously detecting manipulated face images and videos and locating manipulated regions at semantic-level. Our designed multi-branch autoencoder comprises a feature encoder, a shared decoder, four semantic decoders, and a classification network. The output of the feature encoder contains the category information and structure information of the input face image. The structure information is feed into the shared decoder and the four semantic decoders for constructing four different semantic prediction masks. Meanwhile, the category information is feed into the classification network for obtaining semantic prediction labels. Then, the manipulation prediction label and manipulation prediction mask of the input face image can be generated with the semantic prediction masks and semantic prediction labels.

Extensive experiments have been performed to demonstrate the effectiveness of our approach, showing that our approach is capable of accurately detecting and locating manipulated face images and videos, even under cross-manipulation, cross-dataset, and cross-compression scenarios.

**Acknowledgements.** This work was supported by National Key Technology Research and Development Program under 2020AAA0140000.

# References

1. Dang, H., Liu, F., Stehouwer, J., Liu, X., Jain, A.K.: On the detection of digital face manipulation. In: 2020 IEEE/CVF Conference on Computer Vision and Pattern Recognition (CVPR), pp. 5780–5789 (2020). https://doi.org/10.1109/CVPR42600.2020.00582

2. Verdoliva, L.: Media Forensics and DeepFakes: an overview. IEEE J. Sel. Top. Signal Process. **14**(5), 910–932 (2020). https://doi.org/10.1109/JSTSP.2020.3002101

3. Thies, J., Zollhöfer, M., Nießner, M.: Deferred neural rendering: image synthesis using neural textures. ACM Trans. Graph. **38**(4), 1–12 (2019). https://doi.org/10.1145/3306346.3323035. Article 66

4. Thies, J., Zollhöfer, M., Stamminger, M., Theobalt, C., Nießner, M.: Face2Face: real-time face capture and reenactment of RGB videos. In: 2016 IEEE Conference on Computer Vision and Pattern Recognition (CVPR), pp. 2387–2395 (2016). https://doi.org/10.1109/CVPR.2016.262

5. Rössler, A., Cozzolino, D., Verdoliva, L., Riess, C., Thies, J., Niessner, M.: Face-Forensics++: learning to detect manipulated facial images. In: 2019 IEEE/CVF International Conference on Computer Vision (ICCV), pp. 1–11 (2019). https://doi.org/10.1109/ICCV.2019.00009

6. Li, Y., Yang, X., Sun, P., Qi, H., Lyu, S.: Celeb-DF: a large-scale challenging dataset for DeepFake forensics. In: 2020 IEEE/CVF Conference on Computer Vision and Pattern Recognition (CVPR), pp. 3204–3213 (2020). https://doi.org/10.1109/CVPR42600.2020.00327

7. Dolhansky, B., et al.: The deepfake detection challenge (DFDC) dataset. arXiv preprint arXiv:2006.07397 (2020)

8. Li, L., Bao, J., Yang, H., Chen, D., Wen, F.: Advancing high fidelity identity swapping for forgery detection. In: 2020 IEEE/CVF Conference on Computer Vision and Pattern Recognition (CVPR), pp. 5073–5082 (2020). https://doi.org/10.1109/CVPR42600.2020.00512

9. Afchar, D., Nozick, V., Yamagishi, J., Echizen, I.: MesoNet: a compact facial video forgery detection network. In: 2018 IEEE International Workshop on Information Forensics and Security (WIFS), pp. 1–7 (2018). https://doi.org/10.1109/WIFS.2018.8630761

10. Güera, D., Delp, E.J.: Deepfake video detection using recurrent neural networks. In: 2018 15th IEEE International Conference on Advanced Video and Signal Based Surveillance (AVSS), pp. 1–6 (2018). https://doi.org/10.1109/AVSS.2018.8639163

11. Joseph, Z., Nyirenda, C.: Deepfake detection using a two-stream capsule network. In: 2021 IST-Africa Conference (IST-Africa), pp. 1–8 (2021)

12. Li, Y., Chang, M., Lyu, S.: In Ictu Oculi: exposing AI created fake videos by detecting eye blinking. In: 2008 IEEE International Workshop on Information Forensics and Security (WIFS), pp. 1–7 (2018). https://doi.org/10.1109/WIFS.2018.8630787

13. Matern, F., Riess, C., Stamminger, M.: Exploiting visual artifacts to expose deepfakes and face manipulations. In: 2019 IEEE Winter Applications of Computer Vision Workshops (WACVW), pp. 83–92 (2019). https://doi.org/10.1109/WACVW.2019.00020

14. Pan, X., Zhang, X., Lyu, S.: Exposing image splicing with inconsistent local noise variances. In: 2012 IEEE International Conference on Computational Photography (ICCP), pp. 1–10 (2012). https://doi.org/10.1109/ICCPhot.2012.6215223

15. Bappy, J.H., Simons, C., Nataraj, L., Manjunath, B.S., Roy-Chowdhury, A.K.: Hybrid LSTM and encoder-decoder architecture for detection of image forgeries. IEEE Trans. Image Process. **28**(7), 3286–3300 (2019). https://doi.org/10.1109/TIP.2019.2895466

16. Zhou, P., Han, X., Morariu, V.I., Davis, L.S.: Learning rich features for image manipulation detection. In: 2018 IEEE/CVF Conference on Computer Vision and Pattern Recognition, pp. 1053–1061 (2018). https://doi.org/10.1109/CVPR.2018.00116

17. Nguyen, H.H., Fang, F., Yamagishi, J., Echizen, I.: Multi-task learning for detecting and segmenting manipulated facial images and videos. In: 2019 IEEE 10th International Conference on Biometrics Theory, Applications and Systems (BTAS), pp. 1–8 (2019). https://doi.org/10.1109/BTAS46853.2019.9185974

18. Li, L., et al.: Face X-ray for more general face forgery detection. In: 2020 IEEE/CVF Conference on Computer Vision and Pattern Recognition (CVPR), pp. 5000–5009 (2020). https://doi.org/10.1109/CVPR42600.2020.00505

19. Huang, Y., Juefei-Xu, F., Guo, Q., Liu, Y., Pu, G.: FakeLocator: robust localization of GAN-based face manipulations. In: IEEE Transactions on Information Forensics and Security (2022). https://doi.org/10.1109/TIFS.2022.3141262

20. Songsri-in, K., Zafeiriou, S.: Complement face forensic detection and localization with faciallandmarks. arXiv preprint arXiv:1910.05455 (2019)

21. Karras, T., et al.: Progressive growing of GANs for improved quality, stability, and variation. In: International Conference on Learning Representations (2018)

22. Karras, T., Laine, S., Aila, T.: A style-based generator architecture for generative adversarial networks. IEEE Trans. Pattern Anal. Mach. Intell. **43**(12), 4217–4228 (2021). https://doi.org/10.1109/TPAMI.2020.2970919

23. Zhu, J., Park, T., Isola, P., Efros, A.A.: Unpaired image-to-image translation using cycle-consistent adversarial networks. In: 2017 IEEE International Conference on Computer Vision (ICCV), pp. 2242–2251 (2017). https://doi.org/10.1109/ICCV.2017.244

24. Perarnau, G., et al.: Invertible conditional gans for image editing. arXiv preprint arXiv:1611.06355 (2016)

25. Choi, Y., Choi, M., Kim, M., Ha, J.-W., Kim, S., Choo, J.: StarGAN: unified Generative Adversarial Networks for Multi-domain Image-to-Image Translation. In: 2018 IEEE/CVF Conference on Computer Vision and Pattern Recognition, pp. 8789–8797 (2018). https://doi.org/10.1109/CVPR.2018.00916

26. He, Z., Zuo, W., Kan, M., Shan, S., Chen, X.: AttGAN: facial attribute editing by only changing what you want. IEEE Trans. Image Process. **28**(11), 5464–5478 (2019). https://doi.org/10.1109/TIP.2019.2916751

27. Nirkin, Y., Keller, Y., Hassner, T.: FSGAN: subject agnostic face swapping and reenactment. In: 2019 IEEE/CVF International Conference on Computer Vision (ICCV), pp. 7183–7192 (2019). https://doi.org/10.1109/ICCV.2019.00728

28. Li, L., et al.: FaceShifter: towards high fidelity and occlusion aware face swapping. arXiv preprint arXiv:1912.13457 (2019)

29. Yang, X., Li, Y., Lyu, S.: Exposing deep fakes using inconsistent head poses. In: ICASSP 2019–2019 IEEE International Conference on Acoustics, Speech and Signal Processing (ICASSP), pp. 8261–8265 (2019). https://doi.org/10.1109/ICASSP.2019.8683164

30. Li, G., Cao, Y., Zhao, X.: Exploiting facial symmetry to expose deepfakes. In: 2021 IEEE International Conference on Image Processing (ICIP), pp. 3587–3591 (2021). https://doi.org/10.1109/ICIP42928.2021.9506272

31. Li, Y., Lyu, S.: Exposing deepfake videos by detecting face warping artifacts. arXiv preprint arXiv:1811.00656 (2018)

32. Guarnera, L., Giudice, O., Battiato, S.: Deepfake detection by analyzing convolutional traces. In: 2020 IEEE/CVF Conference on Computer Vision and Pattern Recognition Workshops (CVPRW), pp. 2841–2850 (2020). https://doi.org/10.1109/CVPRW50498.2020.00341

33. Zhao, H., Wei, T., Zhou, W., Zhang, W., Chen, D., Yu, N.: Multi-attentional deepfake detection. In: 2021 IEEE/CVF Conference on Computer Vision and Pattern Recognition (CVPR), pp. 2185–2194 (2021). https://doi.org/10.1109/CVPR46437.2021.00222

34. Amerini, I., Galteri, L., Caldelli, R., Del Bimbo, A.: Deepfake video detection through optical flow based CNN. In: 2019 IEEE/CVF International Conference on Computer Vision Workshop (ICCVW), pp. 1205–1207 (2019). https://doi.org/10.1109/ICCVW.2019.00152

35. Wu, X., Xie, Z., Gao, Y., Xiao, Y.: SSTNet: detecting manipulated faces through spatial, steganalysis and temporal features. In: ICASSP 2020–2020 IEEE International Conference on Acoustics, Speech and Signal Processing (ICASSP), pp. 2952–2956 (2020). https://doi.org/10.1109/ICASSP40776.2020.9053969

# Deep Learning Image Age Approximation - What is More Relevant: Image Content or Age Information?

Robert Jöchl[(✉)] and Andreas Uhl

Department of Artificial Intelligence and Human Interfaces,
University of Salzburg, Salzburg, Austria
{robert.joechl,andreas.uhl}@plus.ac.at

**Abstract.** The field of temporal image forensics is the science of approximating the age of a digital image relative to images from the same device. For this purpose, classical methods exist that exploit handcrafted features based on hidden age traces (*i.e.*, in-field sensor defects). In contrast to these classical methods, a Convolutional Neural Network (CNN) learns the features used independently. This has the benefit that other (unknown) age traces can be exploited. However, this also carries the risk of learning non-age-related features to predict the age class. In this work, we analyze the features learned by a standard CNN trained in the context of image age approximation on regular scene images. To analyze whether the model learned to exploit hidden age traces or just other general, non-age-related intra-class properties (*e.g.*, common scene properties or lighting conditions), we applied methods from the field of Explainable Artificial Intelligence (XAI). This analysis is performed with 14 models trained with images from 14 different devices from two datasets.

**Keywords:** Image age approximation · Age features · Image content · Deep learning · Image forensics

## 1 Introduction

The main objective in temporal image forensics is to establish a chronological sequence among pieces of evidence. Since the time-stamp stored in the EXIF header is easy to manipulate, this chronological order must be determined using time-dependent traces hidden in a digital image, like in-field sensor defects. In-field sensor defects are single pixel defects that develop in-field (*i.e.*, after the manufacturing process). Since these defects accumulate over time, the age of a digital image can be approximated based on the detected defects. In principle, these defects affect a single pixel only. However, because of preprocessing like demosaicing (interpolation), the defect spreads to neighbouring pixels. Further characteristics of in-field sensor defects have been studied in multiple publications (*e.g.*, [8, 9, 26–28, 34, 35]).

© The Author(s), under exclusive license to Springer Nature Switzerland AG 2023
X. Zhao et al. (Eds.): IWDW 2022, LNCS 13825, pp. 114–128, 2023.
https://doi.org/10.1007/978-3-031-25115-3_8

In temporal image forensics, it is assumed that a forensics analyst is provided with a set of chronologically ordered trusted images and a second not trustworthy set (from the same device). The goal is to approximate the age of images from the not trustworthy set relative to the trusted set. For this purpose, Fridrich and Goljan proposed a maximum likelihood approach based on median filter residuals in [14]. Since the median filter is a denoising filter and in-field sensor defects appear as image noise, the median filter residuals contain the defect magnitude.

We consider image age approximation as a multi-class classification problem (*i.e.,* where the classes are defined by the different defect onset times and the available trusted images). For this reason, in [21] we propose to utilize traditional machine learning techniques (*i.e.,* a 'Naive Bayes Classifier' and a 'Support Vector Machine') for approximating the image age. Since the age approximation is also based on the presence of in-field sensor defects, a limitation of both approaches [14,21] is that the defect locations must be known beforehand. For this purpose, several methods exist that detect sensor defects based on regular scene images (*e.g.,* [7,11–13,15,27,33]). A defect detection method specifically proposed in the context of image approximation was introduced in [20]. Ahmed et al. proposed another machine learning approach based on in-field sensor defects in [3]. In their work, the defect identification and age approximation are combined in one method.

Unlike these traditional techniques, a Convolutional Neural Network (CNN) learns the classification features used. In this context, Ahmed et al. utilized two well-known CNN architectures for image age approximation (*i.e.,* the AlexNet [24] and GoogLeNet [32]) in [1]. The authors reported an accuracy of over 85% for a five-class classification problem achieved by AlexNet with transfer learning mode. In [19], we systematically investigated the influence of the presence of strong in-field sensor defects when training a CNN for image age approximation. Considering the investigated 'five-crop-fusion' scenario (where five networks are trained on different fixed image patches each) the presence of a strong in-field sensor defect is irrelevant for improving the age classification accuracy. For this reason, we suggested that other 'age' traces are exploited by the network.

The advantage of a CNN learning features independently also carries the risk of learning features that are unrelated to temporal difference between the classes. Deep neural networks can be considered as a 'black box'. For example, in the context of deep learning age approximation, it is not evident that the age class prediction is based solely on detected age traces. In principle, it is likely that images taken in close temporal proximity (*e.g.,* belonging to the same age class) share some common features. For example: (i) common scene properties (*e.g.,* urban or nature scenes); (ii) common weather conditions (*e.g.,* cloudy or blue sky); (iii) seasonal commonalities (*e.g.,* light conditions and vegetation). Such non-age related features can be exploited by the CNN to discriminate between the age classes. An indication that non-age related features are learned was given in [22]. In this work, we investigated if the learned 'age' features are device (in)dependent. However, based on the results obtained, this question could not be answered. In contrast, the results suggest that not solely age related features are learned.

In this work, the main contribution is to analyze the features learned by a standard CNN trained in the context of image age approximation on regular scene images. For this purpose, methods from the field of Explainable Artificial Intelligence (XAI) are applied to investigate whether age features or other non-age-related features (*e.g.* common scene properties or lighting conditions) are learned. The field of XAI is focused on a better understanding and interpretation of the features learned. A relatively recent survey of methods developed in this field is provided in [29]. We utilize two different Class Activation Maps (CAM) techniques to analyse the features learned. In particular, GradCAM++ and ScoreCAM are applied to the same trained models that are also used in [22] to evaluate the device independence. Each model was trained on images from a specific device. In total, images from 14 different devices were available. 10 of these 14 devices are from the publicly available 'Northumbria Temporal Image Forensics' [2] dataset.

The remainder of this paper is organized as follows: the utilized network architecture is described in Sect. 2. In Sect. 3, we give an overview of the used dataset and training settings. The applied CAM techniques (*i.e.*, GradCAM++ and ScoreCAM) are described in Sect. 4 and the results of the CAM analysis are presented in Sect. 5. Potential solutions are outlined in Sect. 6 and the key insights are summarized in the last Sect. 7.

## 2    Steganalysis Residual Network (SRNet)

In [19,22], the idea is that the age traces can be interpreted as a signal that is hidden in a digital image. The field of image steganalysis is the science of detecting whether there is a secret message (signal) hidden in an image. For this reason, methods proposed in the field of image steganalysis may also be suitable for detecting existing age traces. In [19,22], the Steganalysis Residual Network (SRNet), introduced by Boroumand et al. in [6], was used to approximate the age of an image. The SRNet is based on the residual learning principle [18] and depicted in Fig. 1.

The key part of the SRNet is the first seven layers, where no pooling operation is involved. Since pooling act as a low-pass filter, omitting it does not suppress the noise-like stego (age) signal. In [19], the SRNet was trained based on several learning scenarios to investigate the influence of the presence of strong in-field sensor defects and to evaluate if the learned 'age' features are positionally invariant. The 'five-crop-fusion' learning scenario was the most position dependent scenario. This scenario was also utilized in [22] to evaluate the device (in)dependence of the learned features and is again used in this work. In particular, five different SRNets are trained on five different fixed image patches ($256 \times 256$). Each image patch is always extracted from the same location. These locations are at the top left ('tl'), the top right ('tr'), the bottom left ('bl'), the bottom right ('br') corners and in the center of the image ('ce'). The final class

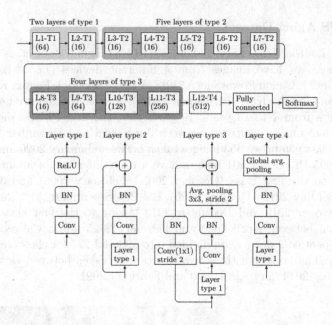

**Fig. 1.** Overview of the SRNet [6].

prediction is obtained by fusing together the different outputs of the individual models.

## 3    Dataset and Training Settings

The SRNet is trained with images of a specific device. In total, images of 14 different devices from two different datasets are available. For each device, a binary classification problem is considered (with a reasonable time interval between the acquisition times of the two classes). Supervised learning is performed based on the binary cross entropy loss. 'AdaMax' [23] is utilized as optimizer with an initial learning rate of 0.001, which is reduced to 0.0001 after 60 epochs. In total, the SRNet is trained over 80 epochs with a batch size of 4. The class with fewer samples is oversampled during training. For each device, the training is performed 10 times, randomly drawing 90% of the available images per class (*i.e.*, stratified). The remaining 10% of the images are used to evaluate the age approximation performance according to the classification accuracy, *i.e.*,

$$\mathrm{acc} = \frac{1}{N} \sum_{i}^{N} I[\hat{y} = y]. \tag{1}$$

$I$ denotes the indicator function that returns a 1 only if the argument is true (*i.e.*, only if the predicted class label $\hat{y}$ is equal to the true class label), and $N$ represents the total amount of test samples. Hence, the classification accuracy is the ratio of correctly predicted test samples.

### 3.1   PLUS Aging Dataset

The Paris Lodron University Salzburg (PLUS) aging dataset is our own dataset where we have images from 4 different devices (*i.e.*, Nikon E7600 (PLUS-nikon01), Canon PowerShotA720IS (PLUS-canon01), Pentax K5 (PLUS-pentax01) and Pentax K5II (PLUS-pentax02)). The time difference between the classes ranges from 7 to 13 years. In particular, the first PLUS-canon01 class consists of 642 images captured between June 2008 and December 2008, and the second class comprises 353 images taken between January 2020 and October 2020. The 995 PLUS-nikon01 images have a time difference of about 14 years; 382 belong to the first class (October 2005 - February 2006) and 350 to the second class (July 2019 - October 2020). For PLUS-pentax01, 316 images taken between October 2013 and December 2013 belong to the first class, and 343 images taken between April 2020 and February 2021 are in the second class. The PLUS-pentax02 images are divided into 598 and 227 for class one and two, respectively. The images of these two classes were taken between October 2014 and December 2014, and April 2020 and February 2021.

(a) PLUS-canon01          (b) PLUS-nikon01

(c) PLUS-pentax01          (d) PLUS-pentax02

**Fig. 2.** Samples of the PLUS aging dataset, the red rectangles represent the five cropping regions and the red crosses the strong in-field sensor defects locations [19] (Color figure online).

All images from all four devices are JPEG compressed RGB color images containing regular scenes (*e.g.*, vacation scenes). Samples of the captured scenes are illustrated in Fig. 2. The PLUS-pentax01 and PLUS-pentax02 were originally available in raw format and JPEG compression was performed with the same fixed settings. Additionally, we are not aware of any changes between the two classes that affect the image acquisition pipeline.

The red rectangles in Fig. 2 represent the five cropping regions, and the defect locations (of all strong in-field sensor defects that have developed between the age classes considered) are illustrated by red crosses. In summary, one out of five image patches contains a defect with the PLUS-nikon01 and PLUS-pentax01, three out of five contain a defect with the PLUS-canon01, and with the PLUS-pentax02, no image patch contains a defect. We remind that, as reported in [19], the presence of strong in-field sensor defects in an image patch is irrelevant for improving classification accuracy in the 'five-crop-fusion' learning scenario.

## 3.2  Northumbria Temporal Image Forensics Database

The Northumbria Temporal Image Forensics (NTIF) [2] database is a publicly available dataset for temporal image forensics. The NTIF dataset comprises images from 10 different cameras (from 5 different models). For each device, approximately 71 time-slots ranging over 94 weeks (between 2014 and 2016) are available. Overall, this results in 41,684 natural color images of indoor and outdoor scenes along with 980 blue-sky scenes. Further details about the dataset can be found in [2]. Images from this database are also used in other image forensics related publications, *i.e.*, [4, 25].

In the context of image age approximation, the NTIF database is used in [1, 3]. The authors in [1] divided the first 25 time-slots into five age classes. As already mentioned in the introduction (Sect. 1), an overall classification accuracy of over 85% is reported for these five classes (by AlexNet with transfer learning mode). In [3] the first 40 time-slots are grouped into five classes. The authors proposed a machine learning approach based on defective pixels and achieved an overall classification accuracy of up to 93%. In this work, the same class definitions as in [1] are used. However, we consider only the first and last classes (*i.e.*, time-slot 1-5 and 21-25, respectively). Overall, there is a time difference of about 4 months between the two classes.

## 4  Explainable Artificial Intelligence (XAI)

Deep neural networks independently learn the features used and are often superior to classical methods based on handcrafted features. However, such networks usually consist of millions of parameters, which turn such models into a 'black box'. The field of XAI focuses on understanding and interpreting the decisions of such deep neural networks. A comprehensive survey of methods in the field of XAI is given in [29].

One group of XAI methods are the so-called Class Activation Maps (CAMs). The original CAM approach was introduced in [38] and indicates the discriminative region used by a CNN to identify a certain class. In principle, the weights of the final output (classification) layer are projected back on the feature maps of the last convolutional layer. The resulting CAM is then upsampled to the size of the input image. CAMs are saliency maps (*i.e.*, topographical representations that highlight the impact of different image regions) and usually visualized as heatmaps.

In this work, two different CAM versions (*i.e.*, the GradCAM++ and Score-CAM) are used. We consider two different CAM versions because the results are not always unambiguous. When analyzing the CAM results in Sect. 5, only those saliency maps are considered where both outputs are similar.

## 4.1  GradCAM++

GradCAM++ was introduced by Chattopadhay et al. in [10] and is an enhanced version of GradCAM [30]. GradCAM is a generalization of CAM, since Grad-CAM has no limitations in the CNN architecture. As the name suggests, Grad-CAM is based on the class-specific gradient information. In particular, an input image is forward propagated through the model. Except for the gradients of the target class, all other gradients are set to zero. These class-specific gradients are backpropagated to the convolutional layer of interest, and the feature maps are weighted according to the obtained gradients. To get a high resolution saliency map (*i.e.*, with the resolution of the input image), the obtained saliency map is combined with guided backpropagation [31].

GradCAM cannot localize multiple occurrences of an object of the same class, or the localization corresponds only to bits and parts of an object. One reason for this is that GradCAM divides the weights (which capture the importance of a particular feature map) by the size of the feature map, *i.e.*, if the response is small or the area of the response is small, the weights decrease. To solve this problem, GradCAM++ uses a more sophisticated backpropagation. To generate the saliency maps for this work, the GradCAM++ implementation provided in [16] is used.

## 4.2  ScoreCAM

In contrast to GradCAM++, ScoreCAM [37] does not rely on gradient information. The authors suggest that the propagating gradients are unstable and generate random noise. To obtain the saliency map, first an image is propagated through the network and the $k$ feature maps from the last convolutional layer are extracted. The obtained feature maps are upsampled (using bilinear-interpolation) to the input image size. After the upsampling, the feature map values are normalized in the range $[0, 1]$. Then, each normalized feature map is multiplied by the input image. This results in $k$ masked input images. All the masked input images are forward propagated through the CNN and the SoftMax class scores (weights) are computed. The final saliency map is the rectified sum of a linear combination between the target class score and each feature map.

In other words, the obtained scores (weights) from forward propagating the masked input images reflect the importance of the feature map. To generate the saliency maps for this work, the ScoreCAM implementation provided in [16] is used.

## 5 CAM Analysis

Image age approximation is a multi-class classification problem, where a class is defined by the temporal resolution of the exploited age traces and the available trusted images. For this purpose, let $\theta$ be the sum of age traces at a certain point in time that are embedded in an image $I$ (e.g., $I + \theta$). Furthermore, we assume that $\theta$ is constant across all images of a given age class $y$ and differs between the other age classes. To approximate the age of an image, the goal of a classifier is to predict the age class $y$ with $\theta_y$ having the highest probability of being embedded in an input image $I$.

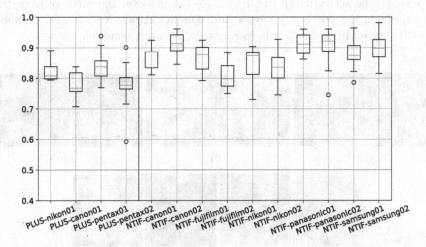

**Fig. 3.** Boxplot of the resulting age approximation accuracy for all 10 runs.

Based on this definition, we assume that if the model can successfully discriminate between age classes, it will learn to detect the age signal $\theta_y$ for a given age class $y$. In Fig. 3, the achieved age classification accuracy for each device and all 10 runs is visualized by a boxplot. We can observe that the learned models can successfully discriminate between the age classes. This would imply that the models learned to detect $\theta$. Based on this assumption, we expect the obtained saliency maps to highlight regions that:

1. are independent of the image content (e.g., captured objects and scene properties),
2. are constant across the different runs (i.e., since all images per class $y$ share the same $\theta_y$, the overall activations should be similar across all different test sets).

Due to space constraints, only some examples of the 35000 generated saliency maps can be shown. However, all generated saliency maps are available under https://wavelab.at/sources/Joechl22b/.

## 5.1  Activation on Objects

Examples of generated saliency maps where the activation is connected to objects
are illustrated in Fig. 4. In particular, we can observe an activation on the head
of a swan (Fig. 4(a)), on the blossoms of multiple flowers (Fig. 4(b)), on a house
(Fig. 4(c)), on two sticks lying on the ground (Fig. 4(d)) and on a rain gutter
(Fig. 4(e)). As already described, the embedded age signal $\theta$ should be invariant
to the presence of objects. An activation solely on objects suggests image content
is likely to be more relevant than the hidden age signal $\theta$ in predicting the age
class. However, one could argue that possibly the color range of an object benefits
the detection of the hidden age signal in this image region. This is contradicted
because: (i) the activation in Fig. 4(b) & (e) extends over the entire objects, (ii)
the objects, in Fig. 4(a) & (c), have a more complex structure with several colors,
(iii) that the entire image patch, in Fig. 4(d), is in the same color range.

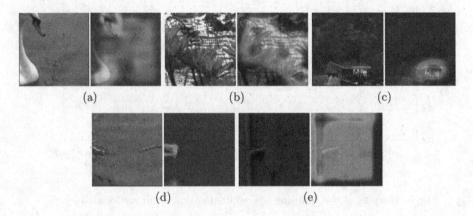

**Fig. 4.** Example of activations directly on objects, the color indicates the importance
of a region, red being very important. (Color figure online)

Another indication that image content is likely to be more important than
the hidden age signal for predicting the age class is shown in Fig. 5. In Fig. 5,
examples of activations on shrub-, tree-like structures are illustrated. Activations
on such structures could be caused when images of one age class contain more
nature scenes than those of the other class. All five examples (Fig. 5(a)–(e)) were
based on images from different NTIF devices, and the respective model predicted
the second age class for all of them.

## 5.2  Activation on Areas

The analysis of the CAMs also revealed that many activations are found in
large homogeneous areas (*e.g.*, sky regions). Examples of such activations are
illustrated in Fig. 6. For example, depending on the weather conditions, the sky

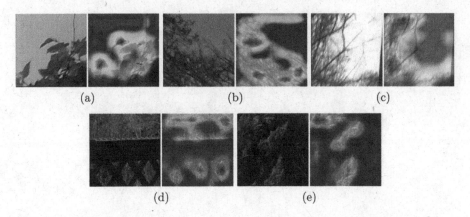

**Fig. 5.** Examples of activations on shrub-, tree-like structures, the color indicates the importance of a region, red being very important. (Color figure online)

can be cloudy (*i.e.*, very bright) or naturally blue. If the sky was cloudy when images of the first age class were taken, and blue when the second age class images were captured, it would be easy for a CNN to discriminate such images based on the sky regions.

To show that some models learned to discriminate between colors, we generated six very bright image patches (*i.e.,* with different shades of white, RGB values from [250, 250, 250] to [255, 255, 255]) and six images with different shades of blue (*i.e.,* RGB values from [0, 0, 250] to [0, 0, 255]). Since these images are generated, there is no age signal embedded. If the model has indeed learned to detect only the embedded age signals (of the different age classes), we would expect the model to select one age class for all generated images. What we do not expect, however, is that the model will assign all images 100% correctly (separate bright and blue images). On average across all 14 devices, these images can be assigned 100% correctly in 51% of the 10 runs when the model is trained with image patches from 'tl', and in 46% of the runs when image patches from 'tr' are used for training. The other results are 37%, 39% and 37% when the model is trained on patches from 'bl', 'br' and 'ce', respectively. The significantly higher results when the model is trained on images patches of both top cropping positions are reasonable, since sky regions are often located in these image regions.

## 5.3   No Constant Activation Pattern

As defined above, we assume that the age signal $\theta$ is constant across all images of a given age class (independent of the train and test set sampling). If the model has learned to detect $\theta$, one would expect to observe a common activation pattern (for correctly predicted samples) across the runs (different train and test sets). To analyze this, the activations for each correctly predicted image of a given run and class are superimposed. In Fig. 7, the superimposed activations of all

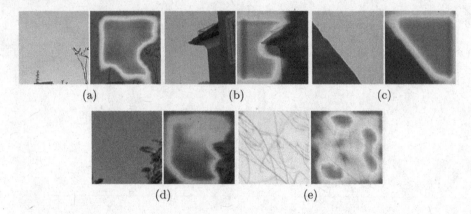

**Fig. 6.** Examples of activations on image areas, the color indicates the importance of a region, red being very important. (Color figure online)

four analyzed runs and the second age class of image patches 'br' of the PLUS-nikon01 images are illustrated. As can be seen, there is no constant activation pattern visible. This is an example, however, for all other image patches and devices, similar results can be observed.

**Fig. 7.** Examples of superimposed activations of correctly predicted image patches of a given run. In this example, the superimposed activations of test images from the PLUS-nikon01, image patch 'br' and the second age class are shown. SubFig. (a) represents the activations of the first run and SubFigs. (b)–(d) the runs 2–4, respectively.

Since instead of the age signal, the image content varies across the different runs, this observation is again an indication that the image content is more relevant for classification than the existing age signal.

## 6  Potential Solutions

Based on the conducted analysis, it is most likely that image content is more important for age classification than the embedded age signal. This is reasonable because the embedded age signal is usually a very weak signal and can easily be overlaid by other, non-age-related inter-class differences (*i.e.,* image content). One potential solution would, therefore, be suppressing the influence of the image content.

This could be achieved by focusing the network on in-field sensor defect locations. In [19], this is done by training the network on small image patches (*i.e.,* $32 \times 32$) extra around each strong in-field sensor defect. However, a disadvantage of this method is that the defect positions must be known in advance. Another approach to suppress the image content could be to apply preprocessing steps, such as median filtering, and train the network with the obtained residuals. The median filter is a denoising filter. Since in-field sensor defects appear as image noise, their presence is preserved in the median filter residuals while most of the image content is suppressed. A further approach could be the utilization of special network architectures. For example, Bayar and Stamm proposed a content suppression layer, called constrained convolutional layer, in [5]. Such a constrained CNN is used in [36] for video camera identification from sensor pattern noise.

Applying the above mentioned constraints (*i.e.,* to the attention region, input data, or network architecture) restricts the feature space. This is intentional, but results in limiting the ability to learn other (unknown) age features as well. For this reason, another approach could be to apply constraints on the acquisition of training data (*i.e.,* to avoid the distraction of common, non-age-related, intra-class properties like common light conditions or scene properties). This could be achieved by capturing standardized scenes, as was done for creating the Dresden DB [17]. However, in the context of image age approximation, the same standardized scenes should be captured for each time-slot. In fact, we are currently creating such a dataset. Potential scene or environmental dependencies are eliminated by capturing different fixed backgrounds and foreground objects in a controlled environment.

## 7   Conclusion

Image age approximation is a multi-class classification problem. Training a CNN in this context has the benefit of independently learning the features used and thus potentially exploiting age features other than the known ones (*i.e.,* in-field sensor defects). However, this also carries the risk of learning other, non-age related features.

In this work, we analyzed the features learned by a standard CNN trained in the context of image age approximation on regular scene images. Based on the investigated CAMs (*i.e.,* GradCAM++ and ScoreCAM) we showed that most likely the image content (*e.g.,* scene properties and light conditions) is more important for classification than the embedded age signal. Because of this observation, it is unlikely that a standard CNN trained on regular scene images would exploit solely age-related features to determine the age class. In particular, in the field of image forensics, it is important that the decision is based on comprehensible evidence (*e.g.,* by methods that rely explicitly on the presence of age traces, such as [14,21]). Thus, when using a CNN for image age approximation, it is important to design the setup carefully.

# References

1. Ahmed, F., Khelifi, F., Lawgaly, A., Bouridane, A.: Temporal image forensic analysis for picture dating with deep learning. In: 2020 International Conference on Computing, Electronics Communications Engineering (iCCECE), pp. 109–114 (2020). https://doi.org/10.1109/iCCECE49321.2020.9231160
2. Ahmed, F., Khelifi, F., Lawgaly, A., Bouridane, A.: The 'northumbria temporal image forensics' database: description and analysis. In: 2020 7th International Conference on Control, Decision and Information Technologies (CoDIT), vol. 1, pp. 982–987 (2020). https://doi.org/10.1109/CoDIT49905.2020.9263888
3. Ahmed, F.N., Khelifi, F., Lawgaly, A., Bouridane, A.: A machine learning-based approach for picture acquisition timeslot prediction using defective pixels. Forensic Sci. Int. Digit. Invest. **39**, 301311 (2021)
4. Al-Ani, M., Khelifi, F.: On the SPN estimation in image forensics: a systematic empirical evaluation. IEEE Trans. Inf. Forensics Secur. **12**(5), 1067–1081 (2017). https://doi.org/10.1109/TIFS.2016.2640938
5. Bayar, B., Stamm, M.C.: Constrained convolutional neural networks: a new approach towards general purpose image manipulation detection. IEEE Trans. Inf. Forensics Secur. **13**(11), 2691–2706 (2018). https://doi.org/10.1109/TIFS.2018.2825953
6. Boroumand, M., Chen, M., Fridrich, J.: Deep residual network for steganalysis of digital images. IEEE Trans. Inf. Forensics Secur. **14**(5), 1181–1193 (2018)
7. Chan, C.H.: Dead pixel real-time detection method for image. US Patent 7,589,770, 15 September 2009
8. Chapman, G.H., et al.: Increases in hot pixel development rates for small digital pixel sizes. Electron. Imaging **2016**(12), 1–6 (2016)
9. Chapman, G.H., Thomas, R., Koren, Z., Koren, I.: Empirical formula for rates of hot pixel defects based on pixel size, sensor area, and ISO. In: Widenhorn, R., Dupret, A. (eds.) Sensors, Cameras, and Systems for Industrial and Scientific Applications XIV, vol. 8659, pp. 119–129. International Society for Optics and Photonics, SPIE (2013). https://doi.org/10.1117/12.2005850
10. Chattopadhay, A., Sarkar, A., Howlader, P., Balasubramanian, V.N.: Grad-CAM++: generalized gradient-based visual explanations for deep convolutional networks. In: 2018 IEEE Winter Conference on Applications of Computer Vision (WACV), pp. 839–847. IEEE (2018)
11. Chen, C.W., Cho, C.Y., Sun, Y.F., Chen, T.M., Su, C.L.: Low complexity photo sensor dead pixel detection algorithm. In: 2012 IEEE Asia Pacific Conference on Circuits and Systems, pp. 360–363. IEEE (2012)
12. Cho, C.Y., Chen, T.M., Wang, W.S., Liu, C.N.: Real-time photo sensor dead pixel detection for embedded devices. In: 2011 International Conference on Digital Image Computing: Techniques and Applications, pp. 164–169. IEEE (2011)
13. El-Yamany, N.: Robust defect pixel detection and correction for bayer imaging systems. Electron. Imaging **2017**(15), 46–51 (2017). https://doi.org/10.2352/issn.2470-1173.2017.15.dpmi-088

14. Fridrich, J., Goljan, M.: Determining approximate age of digital images using sensor defects. In: Memon, N.D., Dittmann, J., Alattar, A.M., III, E.J.D. (eds.) Media Watermarking, Security, and Forensics III, vol. 7880, pp. 49–59. International Society for Optics and Photonics, SPIE (2011)

15. Ghosh, S., Froebrich, D., Freitas, A.: Robust autonomous detection of the defective pixels in detectors using a probabilistic technique. Appl. Opt. **47**(36), 6904–6924 (2008)

16. Gildenblat, J.: Pytorch library for cam methods (2021). https://github.com/jacobgil/pytorch-grad-cam. Accessed 07 June 2022

17. Gloe, T.: Die 'dresden image database' für die entwicklung und validierung von methoden der digitalen bildforensik. In: Fischer, S., Maehle, E., Reischuk, R. (eds.) Informatik 2009 - Im Focus das Leben, p. 172. Gesellschaft für Informatik e. V., Bonn (2009)

18. He, K., Zhang, X., Ren, S., Sun, J.: Deep residual learning for image recognition. In: Proceedings of the IEEE Conference on Computer Vision and Pattern Recognition, pp. 770–778 (2016)

19. Joechl, R., Uhl, A.: Apart from in-field sensor defects, are there additional age traces hidden in a digital image? In: 2021 IEEE International Workshop on Information Forensics and Security (WIFS), Montpellier, France, pp. 1–6 (2021). https://doi.org/10.1109/WIFS53200.2021.9648396

20. Joechl, R., Uhl, A.: Identification of in-field sensor defects in the context of image age approximation. In: 2021 IEEE International Conference on Image Processing (ICIP), Anchorage, AK, USA, pp. 3043–3047 (2021). https://doi.org/10.1109/ICIP42928.2021.9506023

21. Jöchl, R., Uhl, A.: A machine learning approach to approximate the age of a digital image. In: Zhao, X., Shi, Y.-Q., Piva, A., Kim, H.J. (eds.) IWDW 2020. LNCS, vol. 12617, pp. 181–195. Springer, Cham (2021). https://doi.org/10.1007/978-3-030-69449-4_14

22. Joechl, R., Uhl, A.: Device (in)dependence of deep learning-based image age approximation. In: 2022 ICPR-Workshop on Artificial Intelligence for Multimedia Forensics and Disinformation Detection, Montreal, Quebec, Canada, pp. 1–14 (2022)

23. Kingma, D.P., Ba, J.: Adam: a method for stochastic optimization (2017)

24. Krizhevsky, A., Sutskever, I., Hinton, G.E.: ImageNet classification with deep convolutional neural networks. Adv. Neural. Inf. Process. Syst. **25**, 1097–1105 (2012)

25. Lawgaly, A., Khelifi, F.: Sensor pattern noise estimation based on improved locally adaptive DCT filtering and weighted averaging for source camera identification and verification. IEEE Trans. Inf. Forensics Secur. **12**(2), 392–404 (2017). https://doi.org/10.1109/TIFS.2016.2620280

26. Leung, J., Chapman, G.H., Koren, I., Koren, Z.: Characterization of gain enhanced in-field defects in digital imagers. In: 2009 24th IEEE International Symposium on Defect and Fault Tolerance in VLSI Systems, pp. 155–163 (2009)

27. Leung, J., Chapman, G.H., Koren, Z., Koren, I.: Statistical identification and analysis of defect development in digital imagers. In: Rodricks, B.G., Süsstrunk, S.E. (eds.) Digital Photography V, vol. 7250, pp. 272–283. International Society for Optics and Photonics, SPIE (2009). https://doi.org/10.1117/12.806109

28. Leung, J., Dudas, J., Chapman, G.H., Koren, I., Koren, Z.: Quantitative analysis of in-field defects in image sensor arrays. In: 22nd IEEE International Symposium on Defect and Fault-Tolerance in VLSI Systems (DFT 2007). IEEE (2007). https://doi.org/10.1109/dft.2007.59

29. Linardatos, P., Papastefanopoulos, V., Kotsiantis, S.: Explainable AI: a review of machine learning interpretability methods. Entropy **23**(1), 18 (2021)
30. Selvaraju, R.R., Cogswell, M., Das, A., Vedantam, R., Parikh, D., Batra, D.: Grad-CAM: visual explanations from deep networks via gradient-based localization. In: Proceedings of the IEEE International Conference on Computer Vision, pp. 618–626 (2017)
31. Springenberg, J.T., Dosovitskiy, A., Brox, T., Riedmiller, M.: Striving for simplicity: the all convolutional net. arXiv preprint arXiv:1412.6806 (2014)
32. Szegedy, C., et al.: Going deeper with convolutions. In: Proceedings of the IEEE Conference on Computer Vision and Pattern Recognition, pp. 1–9 (2015)
33. Tchendjou, G.T., Simeu, E.: Detection, location and concealment of defective pixels in image sensors. IEEE Trans. Emerg. Top. Comput. 1 (2020). https://doi.org/10.1109/tetc.2020.2976807
34. Theuwissen, A.J.: Influence of terrestrial cosmic rays on the reliability of CCD image sensors part 1: experiments at room temperature. IEEE Trans. Electron Devices **54**(12), 3260–3266 (2007)
35. Theuwissen, A.J.: Influence of terrestrial cosmic rays on the reliability of CCD image sensors part 2: experiments at elevated temperature. IEEE Trans. Electron Devices **55**(9), 2324–2328 (2008)
36. Timmerman, D., Bennabhaktula, S., Alegre, E., Azzopardi, G.: Video camera identification from sensor pattern noise with a constrained convnet. arXiv preprint arXiv:2012.06277 (2020)
37. Wang, H., et al.: Score-CAM: score-weighted visual explanations for convolutional neural networks. In: Proceedings of the IEEE/CVF Conference on Computer Vision and Pattern Recognition Workshops, pp. 24–25 (2020)
38. Zhou, B., Khosla, A., Lapedriza, A., Oliva, A., Torralba, A.: Learning deep features for discriminative localization. In: Proceedings of the IEEE Conference on Computer Vision and Pattern Recognition, pp. 2921–2929 (2016)

# Watermarking

# Physical Anti-copying Semi-robust Random Watermarking for QR Code

Jiale Chen[1], Li Dong[1,2]([✉]), Rangding Wang[1], Diqun Yan[1], Weiwei Sun[3], and Hang-Yu Fan[3]

[1] Department of Computer Science, Ningbo University, Zhejiang, China
{2111082075,dongli,wangrangding,yandiqun}@nbu.edu.cn
[2] The Key Lab of Mobile Network Application Technology of Zhejiang Province, Zhejiang, China
[3] Alibaba Group, Zhejiang, China
{sunweiwei.sww,hangyu.fhy}@alibaba-inc.com

**Abstract.** Recently, QR code has been applied in anti-counterfeiting scenarios, where a unique QR code is attached for a specific item. However, such a QR code-based anti-counterfeiting solution cannot resolve the physical illegal copying issue. The genuine QR code can be physically replicated by scanning and printing. In this work, we propose a physical anti-copying semi-robust randomly watermarking system for QR code. Specifically, the authentic and counterfeit channels a QR code experiences are investigated first. By exploiting the distortion characteristics between two channels, we devise a randomly watermark embedding system, where the watermark bit is embedded via modulating the relationship between two carefully selected transformed coefficients. Finally, to obtain a valid and recognizable binary QR code image, a random binarization procedure is applied, and the regions originally belonging to the white module are erased. The final resultant watermark appears as *white-dot pattern* resides the black module of QR code, which is robust to the authentic print-scan but fragile to the physically illegal copying. Experimental results demonstrate the effectiveness of the proposed watermarking system. This work makes the first step towards exploring semi-robust watermarking for combating physically illegal copying.

**Keywords:** QR code · Semi-robust watermark · Physical anti-copying

## 1 Introduction

Counterfeiting is a criminal offense that involves the fraudulent production and distribution of an item similar to a genuine product. The production, distribution, and sale of counterfeit items not only defrauds those buying the items but also steals profits from the owners and distributors of the genuine articles. To combat the widespread counterfeiting issue, anti-counterfeiting marks can be attached to the genuine product as accessories or printing on the package surface. Traditional anti-counterfeiting countermeasures including micro-text [2], special color

X. Zhao et al. (Eds.): IWDW 2022, LNCS 13825, pp. 131–146, 2023.
https://doi.org/10.1007/978-3-031-25115-3_9

bar printing [15], thermal ink [8], RFID tags [17] or NFC tags [1]. However, in practice, for common consumers, such countermeasures are still far behind satisfactory
due to the lack of professional anti-counterfeiting detection tools or operations.

Recently, QR code has been widely adopted in anti-counterfeiting because
it is cheap and easy to use. One popular QR code anti-counterfeiting scheme
is the *One Item, One QR code* solution [7]. As shown in Fig. 1, this solution
generates a unique QR code and then prints or pastes it on the authentic product. End-user can use mobile devices to scan and decode the QR code, and then
verify the authenticity of products by through an online anti-counterfeiting system. However, there is a flaw for *One Item, One QR code* solution. Considering
that the authentic QR code is printed and published, malicious counterfeiter can
scan, restore and the print a counterfeit QR code, which can pass the authentication as well. This physically illegal copying (IC) attack violates the unique
QR code for one item principle, and poses great threat to the QR code based
anti-counterfeiting. One approach to mitigate this issue is covering up part or
the entire QR code. Consumers can uncover the QR code for verification after
purchase. Unfortunately, this remedy is also flawed because consumers cannot
verify the authenticity before purchase.

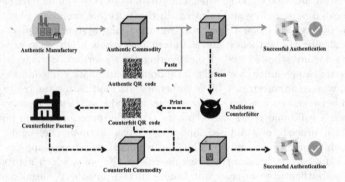

**Fig. 1.** The widely-deployed *One Item, One QR code* anti-counterfeiting solution cannot resist physically illegal copying. The end-user will wrongly authenticate the counterfeiting commodity when the malicious counterfeiter replicates the authentic code by
scanning and printing.

In this work, we propose a semi-robust random QR code watermarking
scheme for solving the physical illegal copying issue. Specifically, the authentic and counterfeit channels a QR code experiences are analyzed, based on
which the watermarking-based physical anti-copying solution is formally formulated. A random embedding is devised, where the watermark bit is embedded
via modulating the relationship of a paired transformed coefficient. We then
apply a random image binarization procedure to obtain a valid binary QR code
image. Finally, the regions originally belonging to the white module are erased,
maintaining the QR code recognition. The final resultant watermark appears
as *white-dot pattern* resides the black module of the QR code, which is robust

over the authentic channel while fragile to the counterfeit channel. Experimental results validate the effectiveness of the proposed watermarking system. The contributions of this work can be summarized as follows,

- We propose a physical anti-copying semi-robust random watermarking scheme for QR codes. For the first time, the semi-robust watermarking technique is introduced for solving the physical illegal copying of QR codes.
- We suggest a transform-domain watermark embedding algorithm and explore its applicability in the semi-robust watermarking context for a discrete binary image.
- A prototype mobile application is developed. Experimental results demonstrate that the proposed watermarking system could achieve authenticity verification of a physical QR code and watermark communication simultaneously.

The rest of this work is organized as follows. Section 2 briefly reviews the related work. In Sect. 3, we analyze and characterize the authentic and counterfeit channels that QR code experiences, based on which Section presents the physical anti-copying semi-robust watermarking system. Experimental results are provided in Sect. 4, and finally, Sect. 5 concludes this work.

## 2   Related Work

### 2.1   Physical Anti-copying

Physical anti-copying (PAC) methods attempt to extract discriminate features that will deviate significantly when a QR code undergoes different communication channels. Pichard *et al.* [12] proposed a dense and random noise pattern, termed Copy Detection Pattern (CDP), for document copying authentication. They then applied CDP to the QR code for product certification [13]. CDP is generated according to the maximum entropy principle, and its high-density randomness ensures its irreversibility. The physically illegal copying makes CDP blurred, which can be easily distinguished from the original CDP. Nguyen *et al.* [11] proposed a reliable performance index of the certification system based on the Neyman Pearson hypothesis test. Recently, Chen *et al.* [4] 2020 proposed a binary classifier-based scheme. The features from both spatial and frequency domains are extracted to train a two-class classifier, which can be used for distinguishing counterfeit barcodes from authentic ones. However, all the aforementioned schemes lack the capability to carry additional watermark bits. To embed data into a QR code, Tkachenko *et al.* [16] proposed a Two-Level QR code. This method replaces the black module of the QR code with a specially designed texture module that encodes data. Thus, the generated QR code can be divided into public and private levels. The standard QR code decoder can be used for the public level to decode it. The private data decodes by maximizing the correlation between the texture module and the candidate template texture modules. Although this scheme carries additional data, its texture module, which is sensitive to the printing and scanning process, is empirically designed and has poor transparency.

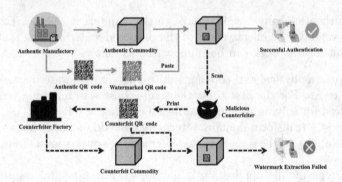

**Fig. 2.** Application scenario for the proposed physical anti-copying semi-robust watermarking system.

## 2.2    Watermarking for QR Code

Image watermarking aims at embedding data (*i.e.*, watermark) into the cover image. It has been successfully applied in many fields, such as copyright protection. Conventionally, image watermarking is vastly discussed in the digital world. However, QR code is often printed and entered into the physical world, then captured and decoded. Thus, when the image watermarking meets the QR codes, one has to consider the robustness of the watermark against printing and capture. There are some robust watermarking schemes developed for resisting printing distortion, *e.g.*, [5,6,10,14]. In addition, some semi-fragile watermarking is only robust to certain types of distortion. Bao *et al.* [3] proposed a watermarking scheme that operates in the transform wavelet domain, which is robust to JPEG compression but sensitive to malicious filtering and random noise. This scheme can be used for image authentication but can not resist printing distortion. Xie *et al.* [19] proposed an anti-counterfeiting watermarking algorithm for QR codes. Still, the watermark can only resist print-and-capture distortion and cannot be against physically illegal copying. In 2021, Xie *et al.* [18] devised an anti-copying 2D barcode by exploiting channel noise characteristics, where the authentication data were stored by exploiting the QR code error-tolerance limit. An authentication decision is made by checking whether the 2D barcode can be correctly decoded. Applying watermarking to physically Illegal Copying (IC) QR codes is quite challenging. As stated in [18], *"... to the best of our knowledge, there is no public report in which a digital watermarking technique has been used against IC attacks."*

In this work, we make the first step towards applying semi-robust randomly watermarking to physically illegal copying. The application scenario is shown in Fig. 2. The watermark bits are embedded into the authentic QR code image and then attached to the package for distribution. This watermark can be correctly extracted when it undergoes an authentic print-and-capture channel, and at the same time, it can not be extracted when the QR code is physically copied. In this next section, we dive into the proposed watermarking system.

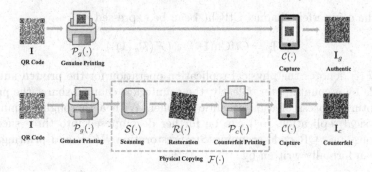

**Fig. 3.** Comparison of the authentic channel and counterfeit channel. Top: The authentic channel (*i.e.*, Print-Capture channel) a QR code experiences. Bottom: The counterfeit channel (*i.e.*, Print-Scan-Print-Capture channel). The key difference lies in additional physical copying action in the counterfeit channel.

## 3 Proposed Physical Anti-copying Watermarking System

In this section, we first investigate the authentic and counterfeit channels that a QR code would undergo, and then model the distortion for these two channels. By exploiting the distortion characteristics between two channels, we propose a physical anti-copying watermarking system.

### 3.1 Model for Authentic and Counterfeit Channels

Remind that this work aims to design an effective semi-robust watermark, which could survive when communicating for the print-then-capture channel (*i.e.*, the authentic channel) while degrading or even invalid for the physical-copying-then-capture (*i.e.*, the counterfeit channel). Therefore, we shall first investigate these two channels and carefully identify and exploit their differences. Based on several previous non-watermark anti-copying schemes [4,9,20], the authentic and counterfeit channels can be modeled as follows.

**Authentic Channel**: As shown in Fig. 3, the authentic channel consists of two critical operations, *i.e.*, printing and scanning, which can be formally expressed as

$$\mathbf{I}_g = \mathrm{AutCh}(\mathbf{I}) \triangleq \mathcal{C}\left(\mathcal{P}_g\left(\mathbf{I}\right)\right), \tag{1}$$

where $\mathrm{AutCh}(\cdot)$ represents the authentic channel, and $\mathbf{I}$ and $\mathbf{I}_g$ are the original digital QR code image, and the captured image by the end-user, respectively. $\mathcal{P}_g(\cdot)$ denotes the genuine printing performed by authentic manufacturer, and $\mathcal{C}(\cdot)$ is the capture process for QR code image. As noted in [9,20], the printing process can be modeled as a linear function, and the capture process can be modeled as low-pass filtering and then re-sampling.

**Counterfeit Channel**: As shown in Fig. 3, a counterfeiter first obtains the printed authentic QR code, and then physically replicate it for fooling consumers.

Thus, the counterfeit channel CtfCh($\cdot$) can be expressed as

$$\mathbf{I}_c = \text{CtfCh}(\mathbf{I}) \triangleq \mathcal{C}\left(\mathcal{F}\left(\mathcal{P}_g\left(\mathbf{I}\right)\right)\right), \tag{2}$$

where $\mathcal{F}(\cdot)$ denotes the physical replication operation for the printed authentic QR code by a counterfeiter. Clearly, the counterfeit channel shares the printing and capture process of the authentic channel. With a thorough examination, the physical replication $\mathcal{F}(\cdot)$ can be further decomposed into three successive operations, $i.e.,$ QR code scanning $\mathcal{S}(\cdot)$, restoration $\mathcal{R}(\cdot)$[1], and printing $\mathcal{P}_c(\cdot)$, which can formally written by

$$\mathcal{F}(\mathbf{I}) = \mathcal{P}_c\left(\mathcal{R}\left(\mathcal{S}(\mathbf{I})\right)\right). \tag{3}$$

The goal of a counterfeiter is to make the counterfeiting physical QR code $\mathbf{I}_c$ as same as possible to the authentic one $\mathbf{I}_g$, $i.e.,$ $\mathbf{I}_c \approx \mathbf{I}_g$. From (1) and (2), one can notice the key difference between authentic and counterfeit channels lies in $\mathcal{F}(\cdot)$. We next analyze the distortion difference between these two channels. First, for a smart counterfeiter, the counterfeiting printing $\mathcal{P}_c(\cdot)$ can be similar to that of the authentic one $\mathcal{P}_g(\cdot)$, by employing similar printing equipment. Second, the aim of the restoration process $\mathcal{R}(\cdot)$ is to mitigate the difference between the captured QR image and the authentic one, using certain restoration techniques such as image binarization. Finally, the scanning operation $\mathcal{S}(\cdot)$ uses a high-resolution scanner (if possible) to scan the physical QR code. Essentially, scanning is a low-pass filtering and re-sampling process similar to the capture process $\mathcal{C}(\cdot)$.

In summary, the dominating distortion over the counterfeit channel stems from the additional scanning operation, suggesting that the distortion of the counterfeit channel suffers additional low-pass filtering and re-sampling distortion. It is worth noting that the distortion of the capture process also incurs low-pass filtering and re-sampling distortion. This requires that an effective physical anti-copying watermarking be semi-robust to low-pass filtering and re-sampling distortion. More specifically, the distortion incurred by an authentic channel requires the anti-copying watermark to be robust. In contrast, the distortion introduced by counterfeit channels requires the anti-copying watermark to be fragile. Thus, we shall carefully design a semi-robust watermarking system, striking the sweet point between fragility and robustness.

Before diving into the proposed watermarking system, we define the physical anti-copying semi-robust watermarking problem formally. Let $\mathbf{w}$ be the watermark bitstream, the watermark embedding process can be expressed as

$$\mathbf{I}^w = \text{Emb}\left(\mathbf{I}, \mathbf{w}\right), \tag{4}$$

where $\text{Emb}(\cdot, \cdot)$ is the watermarking function, embedding watermark $\mathbf{w}$ into the cover QR image $\mathbf{I}$; $\mathbf{I}^w$ is the resultant watermarked image. Upon receiving $\mathbf{I}^w$,

---

[1] The restoration aims at restoring the captured QR code, including denoising, histogram equalization, and binarization $etc.$. This operation is often optional.

the watermark extraction procedure $\text{Ext}(\cdot)$ is performed as follows

$$\mathbf{w} = \text{Ext}\left(\mathbf{I}^w\right). \tag{5}$$

Then, when the watermarked image $\mathbf{I}^w$ communicates over authentic or counterfeit channel, we have

$$\mathbf{w}^g = \text{Ext}\left(\text{AutCh}\left(\mathbf{I}^w\right)\right), \quad \mathbf{w}^c = \text{Ext}\left(\text{CtfCh}\left(\mathbf{I}^w\right)\right), \tag{6}$$

where $\mathbf{w}^g$ and $\mathbf{w}^c$ are the extracted watermark bitstream under authentic or counterfeit channel, respectively. Note that some of the extracted watermark bits would be incorrect. To measure the extraction accuracy, the number of correctly extracted bits can be evaluated by

$$e^g = \sum_i \mathbb{I}\left(w_i^g = w_i\right), \quad e^c = \sum_i \mathbb{I}\left(w_i^c = w_i\right). \tag{7}$$

where $e^g$ and $e^c$ are the number of correctly extracted bits for $\mathbf{w}^g$ and $\mathbf{w}^c$, respectively. $\mathbb{I}(\cdot)$ denotes the indicator function. The goal of the proposed physical anti-copying semi-robust watermarking system are two-fold. First, when the QR code communicates over authentic channel, the extracted watermark shall be *correctly* extracted; and when the QR code communicates over counterfeit channel, the extracted watermark shall be *wrongly* extracted. Thus, the physical anti-copying semi-robust watermarking, consisting of $\text{Emb}(\cdot,\cdot)$ and $\text{Ext}(\cdot)$, should maximize $e^g$ and minimize $e^c$ simultaneously, *i.e.*,

$$\arg\max_{\text{Emb}(\cdot,\cdot),\text{Ext}(\cdot)}\left(e^g - e^c\right). \tag{8}$$

In the next, we present the proposed physical anti-copying semi-robust random watermarking system, attempting to maximize (8).

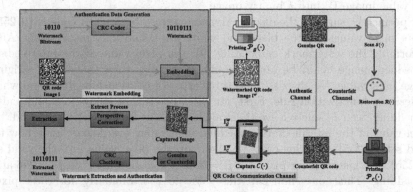

**Fig. 4.** Workflow of the proposed physical anti-copying semi-robust watermarking system.

## 3.2    Watermark Embedding

As shown in Fig. 4, the proposed watermark embedding consists of two steps. First, the authentication checksum data used for error detection is generated and appended to the watermark bitstream. Then, the watermark is embedded into the cover QR code image. Let us first discuss the crucial watermark embedding procedure.

The key idea of the proposed embedding scheme is to design a robustness-controllable embedding algorithm. Note that here the robustness refers to the robustness against low-pass filtering and re-sampling. To implement a robustness-controllable embedding algorithm, we in this work suggest embedding one watermark bit by modulating the relationship between paired transformed coefficients.

More specifically, the original QR code is first divided into overlapping blocks of size $N \times N$. For each image block, the 2D discrete cosine transform (DCT) is then applied, and one can obtain $N \times N$ DCT coefficient matrix $\mathbf{M}$. A pair of coefficients $c_1, c_2$ are selected from the low or middle frequency bands, $e.g.$, $c_1 = \mathbf{M}(12, 19)$ and $c_2 = \mathbf{M}(19, 12)$. Then, the embedding procedure can be formulated as

$$\begin{cases} \hat{c}_1 = \max(c_1, c_2) + \Delta, \hat{c}_2 = \min(c_1, c_2) - \Delta, & \text{if } w = 0 \\ \hat{c}_1 = \min(c_1, c_2) - \Delta, \hat{c}_2 = \max(c_1, c_2) + \Delta, & \text{if } w = 1 \end{cases} \tag{9}$$

where $\hat{c}_1$ and $\hat{c}_2$ are the resultant embedded coefficients. $w \in \{0, 1\}$ is the watermark bit to be embedded, and $\Delta$ is the embedding strength parameter, aiming to enlarge the differences between $\hat{c}_1$ and $\hat{c}_2$. More importantly, $\Delta$ controls the strength of the modification, and thus in fact is the critical parameter to control the robustness. One can obtain the intermediate embedded image $\tilde{\mathbf{I}}^w$ by Inverse-DCT, where each pixel of $\tilde{\mathbf{I}}^w$ is a real value. That is, $\tilde{\mathbf{I}}^w(i, j) \in \mathbb{R}$, where $(i, j)$ are the indices for the $i$-th row and $j$-th column pixel. Considering that a valid QR code shall be a binary-valued image, we need to randomly binarize the real-value image $\tilde{\mathbf{I}}^w$ into a binary image.

Specifically, suppose the discrete dynamic set for binary QR code is $\{0, 255\}$, where pixel values for the black and white are 0 and 255, respectively. After performing the watermark embedding (9), the pixel value of the intermediate embedded image could be larger, equal, or smaller than that of the original image pixel value. We discuss these three types of relationships, based on which to design a binarization rule. First, for these pixels that enjoy no changes after embedding, we can safely leave them alone, without further action. Second, the pixel value of the intermediate embedded image may overflow or underflow the valid set $\{0, 255\}$. For this case, we shall clip the pixel value into the valid set $\{0, 255\}$. For instance, for pixel $\mathbf{I}(i, j) = 255$, it may become $\tilde{\mathbf{I}}^w(i, j) = 256$ after embedding. Thus, one shall clip this value to 255. Similarly, for these pixels $\tilde{\mathbf{I}}^w(i, j) < 0$ whose original pixel values are zeros, one has to clip them to 0. Third, the pixel value of the intermediate embedded image may slightly change, but still, escape the valid set $\{0, 255\}$. For this case, we would like to pull the pixel value to 0 or 255. As a concrete example, suppose the pixel $\mathbf{I}(i, j) = 0$, it

(a) $\mathbf{I}$       (b) $\tilde{\mathbf{I}}^w$       (c) $\mathbf{I}_b^w$       (d) $\mathbf{I}^w$

**Fig. 5.** An example of the embedding process is the randomly embedding strength $p = 0.5$. (a) Original QR code image $\mathbf{I}$. (b) Intermediate embedded QR code image $\tilde{\mathbf{I}}^w$. Note that the $\tilde{\mathbf{I}}^w$ is a real-value watermarked image (normalized in $[0, 255]$ for better visualization), which is quite similar to the original image; zoom in for better comparison. (c) Binarized image $\mathbf{I}_b^w$ according to (10). (d) The final watermarked binary QR image $\mathbf{I}^w$, by erasing the regions that originally belongs to the white module of $\mathbf{I}$.

may become $\tilde{\mathbf{I}}^w(i, j) = 3$ after embedding. Thus, we propose to lift this value to 255 with probability $p$. Similarly, for these pixels $\tilde{\mathbf{I}}^w(i, j) < 255$, whose original pixel values are 255, we suggest downgrading it as 0 with probability $p$. We called $p$ randomly embedding strength. Mathematically, let $\mathbf{I}_b^w = \mathbf{I}$, we summarize the aforementioned binarization operation on the intermediate embedded image $\tilde{\mathbf{I}}^w$ as follows

$$\mathbf{I}_b^w(i, j) = \begin{cases} 255 & \text{if } \tilde{\mathbf{I}}^w(i, j) > 0, \ \mathbf{I}(i, j) = 0, \ p_{ij} \le p \\ 0 & \text{if } \tilde{\mathbf{I}}^w(i, j) < 255, \ \mathbf{I}(i, j) = 255, \ p_{ij} \le p \end{cases} \tag{10}$$

where $p_{ij} \sim U[0, 1]$ and $\mathbf{I}_b^w$ denotes the binarized watermarked image. Finally, to maintain a valid QR code recognition, we propose to erase these regions that are originally belonging to white. This erasion can be expressed by

$$\mathbf{I}^w(i, j) = \begin{cases} \mathbf{I}_b^w(i, j) & \text{if } \mathbf{I}(i, j) = 0 \\ 255 & \text{if } \mathbf{I}(i, j) = 255. \end{cases} \tag{11}$$

In the experiment, we also found that, when the entire QR code image is used for embedding, the QR recognition effectiveness will degrade. This is because the proposed watermark embedding scheme injects specific white-dots into the black module, which could deteriorate the recognition effectiveness of the position detection pattern. To resolve this issue, we suggest excluding the position detection pattern (*i.e.*, the three black squares) and the boundary.

Until now, we obtain the final watermarked QR code image $\mathbf{I}^w$. To intuitively illustrate the proposed embedding procedure, we in Fig. 5 provide an exemplar watermarking process, where the algorithmic parameters are set the same as Sect. 4.1. One can see from Fig. 5-(d) that the semi-robust watermark is rendered as the white-dots in the black module of the QR code. As will be demonstrated shortly, such a white-dot pattern could survive over the authentic channel, while it will be significantly eroded under the counterfeit channel.

Let us go back to the authentication data generation procedure. The goal of this work is to use a watermark to verify the authenticity of the QR code. Thus, we shall verify the correctness of the extracted watermark bits. In this work, we employ the widely-used Cyclic Redundancy Check (CRC) code for checking. Before embedding the watermark into the QR code image, the bitstream encoded with CRC is first obtained as the checksum for the given watermark.

**Remarks**: It is worth noting that the embedding strategy (9) was successfully practiced in several robust watermarking schemes, *e.g.*, [5]. However, none of these works explored the applicability of (9) in the semi-robust watermarking context for a binary image such as QR code.

### 3.3 Watermark Extraction and Authentication

As illustrated in Fig. 4, the general watermark extraction and authentication procedure contains three steps. First, the captured QR code image is prescriptively-corrected, and then the watermark is extracted. Finally, the extracted data is verified through CRC checking.

First, QR code recognition is performed. The standard QR code recognition algorithm includes scanning, image binarization, perspective, geometric correction, and decoding *et al.*. Due to the error-tolerance design of the QR code, the incurred distortion by the embedded watermark does not interfere with the decoding procedure. However, one shall successfully locate and perspective-correct the captured QR code image to facilitate the watermark extraction. Luckily, many off-the-shelf QR codecs are equipped with an efficient automatic positioning and correction algorithm. Thus, in experiments, we employ the QR code localization procedure `QRCodeDetector` provided by OpenCV for locating and perspective correction.

After perspective correction, the perspective-corrected QR code image is divided into non-overlapping blocks, similar to the block division of the embedding procedure. Then for each image block, the DCT transform is applied. The DCT coefficient pairs $\tilde{c}_1$ and $\tilde{c}_2$ extracted from the same coefficient bands used in the embedding procedure. The watermark bit can be extracted by

$$\hat{w} = \begin{cases} 1 & \text{if } \tilde{c}_1 \geq \tilde{c}_2 \\ 0 & \text{if } \tilde{c}_1 < \tilde{c}_2 \end{cases}. \tag{12}$$

Upon extracting the watermark bitstream, CRC checking is conducted. If CRC checking passes, the decision for an authentic QR code is made. Otherwise, excessive erroneous watermark bits are extracted, making the decision counterfeit.

## 4    Experimental Result

### 4.1    Experimental Setup

The size of the QR code image is $246 \times 246$ of version 6, and the image block size for embedding one bit is $30 \times 30$. The length of the randomly-generated

(a) Counterfeit QR code          (b) Authentic QR code

**Fig. 6.** Handheld authentication using prototype mobile app. (a) and (b) are the authentication for the physical counterfeiting QR code, and the authentic QR code, respectively. The authentication results notify on the screen.

**Table 1.** Experimental settings for three constructed datasets. Note that it is unnecessary for an authentic manufacturer to use a scanner to replicate QR code; thus, the cell is noted as NaN.

| Producer | Authentic manufacturer | Counterfeiter I | Counterfeiter II |
|---|---|---|---|
| Printer | Brother MFC-T4500DW (all with 1200 dpi) | Brother MFC-T4500DW (all with 1200 dpi) | RICOH Aficio MP 7500 PCL (all with 1200 dpi) |
| Scanner | NaN | Brother MFC-T4500DW (all with 1200 ppi) | Brother MFC-T4500DW (all with 1200 ppi) |
| Camera | HUAWEI Nova 8 Pro | HUAWEI Nova 8 Pro | One Plus 8 Pro |
| Dataset size | 770 | 770 | 770 |
| Print size (cm) | 1.5 | 1.5 | 1.5 |

watermark bitstream is 59 bits. The embedding coefficient pairs used in this experiment are $\mathbf{M}(19,12)$ and $\mathbf{M}(12,19)$. $\Delta = 50$ and random embedding strength $p = 1.0$. Considering that no publicly available physical anti-copying watermarking datasets. This work constructed three datasets, including one authentic QR code dataset and its two counterfeiting QR code counterparts. The printing size of the QR code image is 1.5 cm × 1.5 cm. The detailed experimental equipment settings for these three datasets are tabulated in Table 1, where each dataset contains 770 samples. The printing resolution is 1200 dpi, and the scanning resolution is 1200 PPI, which is the maximum-available setting provided by the tested equipment. In addition, to verify the practical usage of the proposed method, we have developed a prototype mobile application (see Fig. 6). The size of each captured frame is fixed as 786 × 672 for all tested mobile phones.

(a) $\mathbf{I}^w$          (b) $\mathbf{I}_g^w$          (c) $\mathbf{I}_r^w$          (d) $\mathbf{I}_c^w$

**Fig. 7.** Comparison of the authentic QR code with counterfeited QR code. (a) The digital watermarked QR code image $\mathbf{I}^w$, (b) Captured authentic QR code image that expe-'riences authentic PC channel $\mathbf{I}_g^w = \mathrm{AutCh}(\mathbf{I}^w)$. (c) The counterfeited QR code image is obtained by restoring the scanned physical authentic QR code, $\mathbf{I}_r^w = \mathcal{R}\left(\mathcal{S}\left(\mathcal{P}_g\left(\mathbf{I}^w\right)\right)\right)$. (d) Capture the counterfeit QR code image printed by a counterfeiter $\mathbf{I}_c^w = \mathrm{CtfCh}(\mathbf{I}^w)$.

### 4.2  Comparison of the Authentic and Counterfeited QR Code

As shown in Fig. 7-(a)(b), for the authentic channel, the watermarked QR code image $\mathbf{I}^w$ is authentically printed by the Printer Brother MFC-T4500DW and then captured by the mobile camera of the Huawei Nova 8 PRO. For the counterfeit channel, the authentic QR code image is first scanned by Brother MFC-T4500DW under 1200 dpi, and then the counterfeit QR code image is obtained by printing the scanned QR code image with Brother MFC-T4500DW. Note that we here deliberately use the same printing equipment for both authentic manufacturers and counterfeiters. The reason for this setting is to push the counterfeiting ability to the limit, *i.e.*, the counterfeiter could replicate the QR code using the same equipment as the authentic manufacturer.

By carefully observing the four QR code images from Fig. 7 (a) to (b), one can notice that the number of white-dots in the black module of QR code (*i.e.*, the watermark) is decreasing. This suggests that the embedded watermark erodes gradually. Quantitatively, we test the watermark extraction under 10 trials. For the captured authentic QR code image that experiences PC channel, *i.e.*, $\mathbf{I}_g^w = \mathrm{AutCh}(\mathbf{I}^w)$, the watermark can still be extracted in a low erroneous bit level. The average number of erroneous bits is 0.9, meaning that less than 1 bit goes wrong out of a total of 59 watermark bits. In contrast, for the capture of the counterfeit QR code image printed by a counterfeiter, *i.e.*, $\mathbf{I}_c^w = \mathrm{CtfCh}(\mathbf{I}^w)$, the average erroneous bit is 28.9, closing to the 29.5 erroneous bits of the random-guessing watermark extraction.

### 4.3  Printing Size v.s. Erroneous Bits

In general, the printing size of the anti-copying QR code depends on the printing equipment. Considering that the printing resolution of a printer is limited, a QR code can hardly be printed faithfully as its digital version. Thus the watermark cannot be extracted correctly when the printing size is too small. Therefore, finding the relationship between the printing size and the number of erroneous bits is important. To this end, we print different QR codes of various sizes,

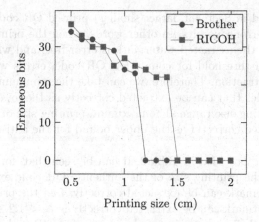

**Fig. 8.** The printing size (cm) versus the number of erroneous bits.

ranging from 0.5 cm × 0.5 cm to 2.5 cm × 2.5 cm, and then record the number of erroneous bits. Each QR code is recognized for 30 attempts. The minimum number of erroneous bits among these attempts is recorded as the final result. As shown in Fig. 8, the number of watermark erroneous bits decreases w.r.t. the increment of the printing size. The printing size of the QR code is negatively correlated with the number of error bits. In this light, one can observe the lower bound of the printing size, where the watermark cannot be correctly extracted for the printing size when smaller than this lower bound.

**Table 2.** The range of test printing size for authentic manufacturer and counterfeiter.

| Printer | Authentic | Counterfeit I | Counterfeit II |
|---|---|---|---|
| Upper bound (cm) | +∞ | 1.8 | 2.3 |
| Lower bound (cm) | 1.2 | 0.0 | 0.0 |

## 4.4   Printing Size v.s. Anti-copying Capability

In practice, physical anti-copying watermarking supports smaller printing size is preferred, which can be attributed to two reasons. First, smaller printed QR code can find more application scenarios, *e.g.*, delicate package. Second, acquiring a high-resolution image for smaller printed QR code is costly, and thus small printing size barriers the counterfeiting; when the QR code printed large enough, it can be forged counterfeited even using a low-resolution scanner or printer. We in this section aim to empirically find the feasible printing size for support reasonably good anti-copying capability of QR code.

To find a feasible range of the printing size, we have printed QR codes of different sizes at an interval of 0.1 cm, and counterfeit them with different printers.

As aforementioned in Sect. 4.3, larger(smaller) printed QR code often leads to fewer(more) the erroneous bits. In other words, when the printing size greater than a threshold, the extracted watermark is error-free; and when the printing size is less than a threshold for counterfeit QR code, errors would occur during watermark extraction. Therefore, we can take the minimum printing size of authentic QR codes that can be extracted correctly as the lower bound for the anti-copying printing size range. The maximum printing size of counterfeit that cannot be extracted correctly as the upper bound for the anti-copying printing size range.

The results are provided in Table 2. It can be seen that, for authentic manufacturer, when the printing size of the authentic QR code exceeds 1.2 cm × 1.2 cm, the watermark can be extracted correctly, $i.e.$, the printing size range of which the watermark can be extracted correctly is $\mathbb{A} = [1.2, +\infty)$. For Counterfeiter I, when the printing size of the counterfeit QR code is less than 1.8 cm × 1.8 cm, the watermark cannot be extracted correctly, $i.e.$, the printing size range the watermark cannot be extracted is $\mathbb{F}_1 = (0, 1.8]$. Similarly, for Counterfeiter II, the printing size range the watermark cannot be extracted is $\mathbb{F}_2 = (0, 2.3]$. Therefore, a feasible printing size range of anti-copying should be $\mathbb{A} \bigcap \mathbb{F}_1 \bigcap \mathbb{F}_2 = [1.2, 1.8]$.

**Table 3.** Performance Comparison with Chen $et\ al.$ [4]. The best results highlighted in bold.

| Method | FAR | FRR | NACC | AUC |
|---|---|---|---|---|
| Chen $et\ al.$ [4] | 0.00% | 2.50% | 98.75% | 0.9958 |
| Proposed | 0.00% | **0.52%** | **99.74%** | **0.9974** |

### 4.5   Comparison of Authentication Performance

To the best of our knowledge, few works realize the physical anti-copying function from the watermarking perspective. To this end, we compare the most recent and relevant work Chen $et\ al.$ [4]. They employed spatial and frequency features to train a two-class classier to distinguish the authentic QR from the counterfeit ones. Instead, we report the authenticity of QR based on the success or failure of semi-robust watermark extraction. False Acceptance Rate (FAR, the percentage of counterfeit samples that have been falsely accepted as authentic), False Rejection Rate (FRR, the percentage of genuine samples that have been falsely accepted as counterfeit), and Normalized ACCuracy (NACC) are employed as performance metrics. The NACC is defined as follows,

$$NACC = 1 - (FAR + FRR)/2 \qquad (13)$$

Experiments were conducted on the datasets shown in Table 1. Experimental results are given in Table 3. Compared with Chen $et\ al.$ [4]. The proposed method always achieves superior performance under all the metrics. Specifically, our proposed semi-robust watermarking solution shows advantages in anti-copying performance, with a higher accuracy rate of 99.74% and lower FRR of 0.52%. Despite the superior performance, our method provides an additional communication channel via semi-robust watermarking, while Chen *et al.* [4] merely made a binary decision without the capability of carrying additional information.

## 5   Conclusion

In this work, we made the first step toward implementing a physical anti-copying semi-robust watermarking for QR codes. We devised a random watermark embedding procedure by exploiting the distortion characteristics between the authentic and counterfeit channels. The resultant semi-watermark appears as irregular *white-dot pattern* resides the black module of QR code, which is robust to the authentic print-scan but fragile to the physically illegal copying. Compared with existing physical anti-copying approaches, the proposed scheme requires no training data to train classifiers. More importantly, the proposed method provides the verification of authenticity for a QR code and additional communication capability for transmitting watermarks simultaneously. Experimental results demonstrate the effectiveness of the proposed watermarking system. We also developed a prototype mobile app to verify the practical usage of the proposed method. We would like to extend the proposed scheme to color barcode cases for future work.

**Acknowledgements.** This work was supported by the National Natural Science Foundation of China (61901237, 62171244), Alibaba Innovative Research Program. Ningbo Natural Science Foundation- Young Doctoral Innovation Research Project (Grant No. 2022J080).

## References

1. Alzahrani, N., Bulusu, N.: Securing pharmaceutical and high-value products against tag reapplication attacks using NFC tags. In: 2016 IEEE International Conference on Smart Computing (SMARTCOMP), pp. 1–6. IEEE (2016)
2. Baldini, G., Fovino, I.N., Satta, R.: Survey of techniques for fight against counterfeit goods and intellectual property rights (IPR) infringing (2015)
3. Bao, P., Ma, X.: Image adaptive watermarking using wavelet domain singular value decomposition. IEEE Trans. Circuits Syst. Video Technol. **15**(1), 96–102 (2005)
4. Chen, C., Li, M., Ferreira, A., Huang, J., Cai, R.: A copy-proof scheme based on the spectral and spatial barcoding channel models. IEEE Trans. Inf. Forensics Secur. **15**, 1056–1071 (2020)
5. Fang, H., Zhang, W., Zhou, H., Cui, H., Yu, N.: Screen-shooting resilient watermarking. IEEE Trans. Inf. Forensics Secur. **14**(6), 1403–1418 (2019)
6. Kang, X., Huang, J., Zeng, W.: Efficient general print-scanning resilient data hiding based on uniform log-polar mapping. IEEE Trans. Inf. Forensics Secur. **5**(1), 1–12 (2010)
7. Keni, H., Earle, M., Min, M.: Product authentication using hash chains and printed QR codes (2017)

8. Lehtonen, M.O., Michahelles, F., Fleisch, E.: Trust and security in RFID-based product authentication systems. IEEE Syst. J. **1**(2), 129–144 (2007)
9. Malvido, A., Pérez-González, F., Cousiño, A.: A novel model for the print-and-capture channel in 2D bar codes. In: Gunsel, B., Jain, A.K., Tekalp, A.M., Sankur, B. (eds.) MRCS 2006. LNCS, vol. 4105, pp. 627–634. Springer, Heidelberg (2006). https://doi.org/10.1007/11848035_83
10. Nakamura, T., Katayama, A., Kitahara, R., Nakazawa, K.: A fast and robust digital watermark detection scheme for cellular phones. NTT Tech. Rev. **4**, 57–63 (2006)
11. Nguyen, H.P., Retraint, F., Morain-Nicolier, F., Delahaies, A.: A watermarking technique to secure printed matrix barcode-application for anti-counterfeit packaging. IEEE Access **7**, 131839–131850 (2019)
12. Picard, J.: Digital authentication with copy-detection patterns. Electron Imaging **5310**, 176–783 (2004)
13. Picard, J., Landry, P., Bolay, M.: Counterfeit detection with QR codes. In: Proceedings of the 21st ACM Symposium on Document Engineering. Association for Computing Machinery, New York, NY, USA (2021). https://doi.org/10.1145/3469096.3474924
14. Pramila, A., Keskinarkaus, A., Seppänen, T.: Toward an interactive poster using digital watermarking and a mobile phone camera. SIViP **6**, 211–222 (2012)
15. Song, B., Mitchell, C.J.: RFID authentication protocol for low-cost tags. In: Proceedings of the first ACM conference on Wireless network security, pp. 140–147 (2008)
16. Tkachenko, I., Puech, W., Destruel, C., Strauss, O., Gaudin, J.M., Guichard, C.: Two-level QR code for private message sharing and document authentication. IEEE Trans. Inf. Forensics Secur. **11**(3), 571–583 (2016)
17. Turcu, C.E., Turcu, C.O., Cerlinca, M., Cerlinca, T., Prodan, R., Popa, V.: An RFID-based system for product authentication. In: Eurocon 2013, pp. 32–39. IEEE (2013)
18. Xie, N., Zhang, Q., Chen, Y., Hu, J., Luo, G., Chen, C.: Low-cost anti-copying 2D barcode by exploiting channel noise characteristics. IEEE Trans. Multimedia **23**, 3752–3767 (2021). https://doi.org/10.1109/TMM.2020.3031083
19. Xie, R., Hong, C., Zhu, S., Tao, D.: Anti-counterfeiting digital watermarking algorithm for printed QR barcode. Neurocomputing **167**, 625–635 (2015)
20. Zhang, L., Chen, C., Mow, W.H.: Accurate modeling and efficient estimation of the print-capture channel with application in barcoding. IEEE Trans. Image Process. **28**(1), 464–478 (2019)

# Robust and Imperceptible Watermarking Scheme for GWAS Data Traceability

Reda Bellafqira[1]([✉]), Musab Al-Ghadi[1], Emmanuelle Genin[2], and Gouenou Coatrieux[1]

[1] IMT Atlantique, Inserm, UMR 1101 LaTIM, Brest, France
{reda.bellafqira,musab.al-ghadi,
gouenou.coatrieux}@imt-atlantique.fr
[2] Inserm UMR 1078, Brest, France
emmanuelle.genin@inserm.fr

**Abstract.** This paper proposes the first robust watermarking method of outsourced or shared genomic data in the context of genome-wide association studies (GWAS) with the primary purpose of identifying the individual or entity at the origin of an illegal information redistribution or disclosure. Our scheme's first unique feature is that it employs a database watermarking strategy to take advantage of the fact that GWAS data are stored in variant call format (VCF) files, which have a database-like structure. Second, it proposes a quantization index modulation based on watermarking modulation for GWAS data under the constraint of not interfering with identifying candidate variants or genes involved in the pathology. We evaluate the theoretical performance of our method in terms of watermarking insertion capacity, distortion, and robustness against different attacks. Experimental results conducted on real data and the weighted-sum statistic (WSS) GWAS study demonstrate the efficiency of the proposed scheme and that it can be used for identifying the cloud service providers (geneticists) at the origin of an information disclosure even if the genotype data has been modified.

**Keywords:** Information security · Genome-wide association studies (GWAS) · Traceability · Watermarking · Genomic data

## 1 Introduction

Nowadays, genomic data are widely collected, stored, processed, and shared for various genomic applications. They can be used in legal and forensics, where a DNA (DeoxyriboNucleic Acid) sample found on a victim or at a crime scene may be exploited as a shred of evidence by law enforcement to track down suspected criminals. In healthcare, genomic data are guiding medical decisions. For instance, it has been demonstrated that women with specific genetic variants in the BRCA (BReast CAncer) genes have about

R. Bellafqira and M. Al-Ghadi—Equal contributions.
This work was supported in part by the French Government support granted to the Labex Comin-Labs and managed by the ANR through the "Investing for the Future" Program under Grant ANR-10-LABX-07-01 through the project TADOP.

an 80% chance of developing breast cancer [1]. Therefore, identification of some individuals who carry these variants can help them to opt for preventive mastectomies [2]. In research, genomic data are being used to discover new associations between traits and some diseases. In this case, association tests are conducted through GWAS, the objective of which is to detect genetic variants that are associated with some complex disorders or diseases [3–6].

In general, a GWAS corresponds to an observational study of a set of genetic variants in the genomes of different individuals in order to see if any variant or a set of variants located in a specific region of the genome (e.g., a set of genes) is associated with a disease [7]. The usual design to conduct a GWAS test is a cases-controls, where genotype distributions at different genetic positions are compared between samples of individuals affected by the disease of interest (cases) and unaffected individuals from the same population (controls). This is the case of the WSS algorithm [8], the objective of which is to compare the number of genotypes in a set of genetic variants from both cases and controls for a studied gene. These association tests are externalized in cloud environments to access high-capacity storage and computation capabilities. However, outsourcing genomic data induces several security issues in confidentiality, traceability, integrity, or traitor tracing, ranging from unintentional disclosure of data due to human errors to planned attacks. Furthermore, genomic data are vulnerable because they allow the owner's unique identification [9]. In this work, we are interested in securing genomic data used in case-control studies such as WSS by watermarking to ensure traitor tracing, i.e., the identification of individuals who are the origin of illegal information disclosure. Different tools have been proposed in order to ensure the security of outsourced data. They are based on various security mechanisms such as encryption [10], digital signatures [11,12], data structure [13] or watermarking [14,15]. Even though digital signature and data structure-based solutions are more commonly used in database management systems to verify integrity, they introduce additional information in the data. Encryption-based methods allow the protection of data confidentiality, but data are no longer protected once they are decrypted. Contrary to all these categories, watermarking relies on the invisible embedding of a message, i.e., a watermark into host data, by imperceptibly modifying them with the constraint that the introduced distortion is controlled. This mechanism leaves access to watermarked data while maintaining them protected. Depending on the relationship between the host data and the embedded watermark, watermarking solutions can be used to achieve different security goals, such as ensuring data integrity, protecting copy rights, or finding spies. These methods were proposed either for using genomic data as a storage mediums [16,17], for protecting messages in genomic data [18,19] or for protecting genomic data themselves [20–22]. They can be used for watermarking the DNA of living organisms or not. All these methods allow genomic data watermarking for various purposes. They were proposed for cellular DNA, and they can not be used for genomic data outsourced for GWAS.

This paper presents the first robust watermarking method for ensuring traitor tracing for genomic data externalized for GWAS studies. Our method is unique in that it employs a database watermarking strategy to capitalize on the fact that GWAS data are stored in Variant Call Format (VCF) files, which have a database-like structure. And, it proposes a quantization index modulation (QIM) [23] based on watermarking modulation for GWAS data under the constraint of not interfering with identifying candidate variants or genes involved in the pathology. The contributions of this paper can be summarized as follows: (i) To the best of our knowledge, the proposed approach is the first attempt to demonstrate the application of watermarking on genomic data, specifically securing the variant genetic sequences stored in the VCF file and used in GWAS. (ii) The watermark is secretly embedded within genomic data without violating genomic processing, such as identifying candidate variants or genes involved in the pathology. (iii) The results show that the proposed approach inserts the watermark in the genomic data with very low data distortion and high robustness to common watermark attacks such as tuples suppression and addition.

The rest of this paper is organized as follows: In Sect. 2, we come back to the introduction to genomic data and database models. Section 3 provides the details of the watermarking solution we propose, while Sect. 4 details the theoretical performance of our solution. Experimental results and discussion are presented in Sect. 5 and conclusions are given in Sect. 6.

## 2    Genomic Data and Database Model

This section briefly introduces genomic data used in GWAS, particularly VCF files and weighted sum statistic (WSS) files, before detailing the database model.

### 2.1    Genomic Data

The human body is made up of billions of cells where each has one nucleus, and this nucleus contains 23 pairs of chromosomes. The complete set of all the DNA contained in one cell is called the genome, and the basic unit of heredity is a particular part of the genome called a gene. Human beings have all the same number of genes, each controlling a particular behavior. However, some behaviors do not express themselves in the same way. These differences between individuals' genomes are called genetic variations. Genetic variants account for about 1% of the difference between two people. One can distinguish three main types of genetic variants [24]: SNP (Single Nucleotide Polymorphisms) that corresponds to a substitution of a single nucleotide at a specific position in the genome; indels (insertions/deletions) that correspond to the insertion or deletion of several nucleotides in the genome; and structural variants that correspond to the deletions, duplications, or rearrangements of large sections of a chromosome or even whole chromosomes. Genetic variants are common genomic data that are used for performing GWAS, and these data are kept in VCF files [25]. The variant call format was developed in order to standardize large-scale genetic variant sharing and storage. A VCF file corresponds to a text file consisting of three parties: meta-data lines, a header line, and data lines. Meta-data lines that begin the file and are included after ## provide data line descriptions. The header line started by # names the columns for data lines. Finally, data lines follow the header line, and each data line or record represents one variant of a given position in the genome. Among several columns per data line present in the VCF file, eight of them are fixed. These are: **CHROM**: a unique identifier from the reference genome that corresponds to chromosome number. **POS**: refers to the position of first base on the reference genome. **ID**: a unique identifier for each record if it exists. **REF**: reference base(s). **ALT**: alternate base(s). **QUAL**: a measure of the quality of the identification of ALT. **FILTER**: filter status. **INFO**: gives additional information such as the number of individuals, frequency alleles, etc. If genotype data is present, fixed columns are followed by a FORMAT column which specifies the data types and order, then an arbitrary number of genotyped individuals. Notice that the dot '.' symbol represents missing value. As shown in Fig. 1, each column that represents genotyped individuals contains genotype information with the same data type indicated in the FORMAT column. One of the significant components of the genotype information is genotype values (GT) which encodes alleles as numbers separated by '|' or '/'; 0 indicates the reference allele, 1 indicates the first allele listed in ALT, 2 indicates the second allele listed in ALT and so on. Therefore, GT could be $0/0, 0/1, 1/2, ./1$ or $1/1$, etc.

```
##fileformat=VCFv4.3
##reference=file:///seq/references/1000GenomesPilot-NCB136.fasta
##contig=<ID=20,length=62435964,assembly=B36,md5=f126c8ffb2da,taxonomy=x>
##INFO=<ID=DP,Number=1,Type=Integer,Description="Total Depth">
##INFO=<ID=AF,Number=A,Type=Float,Description="Allele Frequency">
##INFO=<ID=DB,Number=0,Type=Flag,Description="dbSNP membership>
##FILTER=<ID=q10,Description="Quality below 10">
##FILTER=<ID=s50,Description="Less than 50% of samples have data">
##FORMAT=<ID=GT,Number=1,Type=String,Description="Genotype">
##FORMAT=<ID=DP,Number=1,Type=Integer,Description="Read Depth">
```

| #CHROM | POS | ID | REF | ALT | QUAL | FILTER | INFO | FORMAT | NA00001 | NA00002 | NA00003 |
|---|---|---|---|---|---|---|---|---|---|---|---|
| 20 | 1234567 | microsat1 | GTC | G,GTCT | 50 | PASS | DP=9 | GT:DP | 0/1:4 | 0/2:2 | 1/1:3 |
| 20 | 17330 | . | T | A | 3 | q10 | DP=11;AF=0.017 | GT:DP | 0/0:3 | 0/1:5 | 0/0:41 |
| 20 | 1110696 | rs6040355 | A | G,T | 67 | PASS | DP=10;AF=0.333,0.667;DB | GT:DP | 0/2:6 | 1/2:0 | 2/2:4 |
| 20 | 1230237 | . | T | . | 47 | PASS | DP=13 | GT:DP | 0/0:7 | 0/0:4 | ./.:. |
| 20 | 1234567 | microsat1 | GTC | G,GTCT | 50 | PASS | DP=9 | GT:DP | 0/1:4 | 0/2:2 | 1/1:3 |

**Fig. 1.** An example of VCF file. It stores genomic data, in particular, genetic variants.

## 2.2 Weighted Sum Statistic (WSS) Method

To conduct GWAS, individuals who are affected (cases) and unaffected (controls) are genotyped to produce thousands or up to millions of genetic variants stored into VCF files. After that, an intermediary step is conducted to generate other files specific to each GWAS. In this paper, we are interested in watermarking WSS files. As illustrated in Table 1, a WSS file is composed of the following columns: CHROM, POS, ID, REF, ALT, and an arbitrary number of individuals. The WSS is GWAS method that was proposed in [8] as a tool for the identification of the association of rare variants with diseases. Some studies have pointed out that groups of multiple rare variants together can explain a large proportion of the genetic basis for some diseases. In WSS, variants are grouped according to their biological functionality (e.g., gene), and each individual is scored by a weighted sum of the variant counts. To test for an excess of variants in affected individuals, we use a permutation of disease status among affected and unaffected individuals. Using permutations, the WSS method adjusts the variant weights and the requirement that a mutation is observed to be included in the study. In WSS, rare variant counts within the same gene for each individual are accumulated rather than collapsing. Then, it introduces a weighting term to emphasize alleles with a low frequency in unaffected individuals. Finally, the scores for all individuals are ordered, and then, WSS is computed as the sum of ranks for cases. A permutation procedure determines the significance based on the p-value.

**Table 1.** An example of VCF file; it stores genomic data, particularly genetic variants.

| #CHROM | POS | ID | REF | ALT | NA00001 | NA00002 | NA00003 |
|--------|-----|-----|-----|-----|---------|---------|---------|
| 20 | 1234567 | microsat1 | GTC | G | 0 | 0 | 1 |
| 20 | 1234567 | rs6040355 | GTC | GTCT | 1 | 2 | 0 |

## 3   Proposed Database Watermarking Scheme for GWAS Data

In this section, we first present the standard chain of database watermarking [26], the QIM and, by next, the watermarking scheme we propose for WSS data. Before entering in detail, we illustrate in the Table 2 the acronyms used in our scheme.

**Table 2.** Acronyms that are used in the watermarking method we propose.

| | | | |
|-----|-----|-----|-----|
| $\Delta$ | Quantization step (distortion factor) | $w$ | A watermark bit $w \in 0,1$ |
| $N_g$ | Number of groups | $S_g$ | Number of tuples for each group |
| $G$ | Group of tuples, i.e. $\{G_i\}_{i=1,\cdots,N_g}$ | $G^A$ | Sub-group $A$ of tuples in $G$ |
| $G^B$ | Sub-group $B$ of tuples in $G$ | $|C_0^A|$ | Cardinality of zero values in sub-group $G^A$ |
| $|C_0^B|$ | Cardinality of zero values in sub-group $G^B$ | $d$ | $|C_0^A| - |C_0^B|$ |
| $N_c$ | Number of columns in the WSS file | $D_\Delta$ | Percentage of modulation for a given $\Delta$ |
| $N_r$ | Number of tuples in the WSS file (i.e. $S_g \times N_g$) | $db_{size}$ | Size of WSS file (i.e. $N_r \times N_c$) |

### 3.1   Database Watermarking

By definition, a database is an organized collection of data that are generally stored and accessed from a computer system. Formally, a database $DB$ refers to a finite set of tables or relations $\{R_i\}_{i=1,\cdots,N_r}$. From hereon and for sake of simplicity, we will consider a database that contains one single relation constituted of $N$ tuples $\{t_u\}_{u=1,\cdots,N}$, each of $M$ attributes $\{A_1, A_2, \cdots, A_M\}$. The attribute $A_n$ takes its values within an attribute domain, and $t_u.A_n$ refers to the value of the $n^{th}$ attribute of the $u^{th}$ tuple of the database. The value $t_u.PK$ is an attribute value or a set of attribute values, represents the unique identifier of each tuple of the database. In the literature, most schemes that have been proposed for database watermarking follow the process illustrated in Fig. 2.

This process is based on two basic procedures: watermark embedding and watermark detection/extraction. The watermark embedding procedure includes a pretreatment, the purpose of which is to make the watermark insertion/extraction independent of the database structure or the way the database's data is stored. To do so, database tuples are grouped into $N_g$ non-overlapping groups $\{G^i\}_{i=1,\cdots,N_g}$. This grouping is usually conducted by calculating the index number $n_u \in [0, N_g - 1]$ of each group for the tuple $t_u$ such that

$$n_u = H(K_w|H(K_w|t_u.PK)) \mod N_g \tag{1}$$

where $H$, $K_w$, and | represent the cryptographic hash function, the secret watermarking key, and the concatenation operator, respectively. We use a cryptographic hash function, such as the Secure Hash Algorithm (SHA), to ensure certain grouping and equal distribution of tuples into different groups. After database partitioning, one bit of the watermark is inserted into each group of tuples by modifying or modulating attribute values accordingly to the rules of the retained watermarking modulation, such as the order of database tuples [27]. Therefore, within a database of $N_g$ groups, a watermark $W = \{w_i\}_{i=1,\dots,N_g}$ of $N_g$ bits is embedded. The watermark detection works similarly. First, the database is partitioned into $N_g$ groups based on the secret watermarking key $K_w$. Then, one watermark bit is extracted or detected from each group based on the used modulation. In the sequel, we explain the proposed method, which follows these procedures and is based on QIM and majority vote.

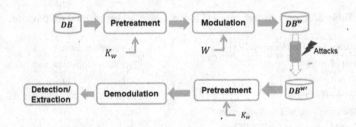

**Fig. 2.** A common database watermarking chain.

## 3.2 Quantization Index Modulation (QIM)

QIM [28] relies on quantifying the host data (e.g., image, database) components by rounding each component to the nearest odd/even quantized value according to the value of the watermark bit $w$ and a quantization step size $\Delta$. More specifically, let $w \in \{0,1\}$ be a watermark bit, $\Delta$ be a quantization step size that controls the level of distortion. In the QIM method, according to the value of the watermark bit to be embedded, the host data components are shifted by $\pm\Delta$. In this work, we apply this QIM method in order to embed one watermark bit $w_i$ into each group of tuples, i.e. $\{G^i\}_{i=1,\cdots,N_g}$. More clearly, let $w_i \in \{0,1\}$ be a watermark bit, $\Delta$ be a quantization step size that control the level of distortion and $d$ be the difference between the cardinality of zero values in sub-group $G_A$ ($|C_0^A|$) and sub-group $G_B$ ($|C_0^B|$) for each individual ($P_i$: $i = 1, \cdots, |\text{patients}|$), where

$$d = |C_0^A|_{P_i} - |C_0^B|_{P_i} \tag{2}$$

According to the value of $w_i$, $d$ is rounded to the nearest even/odd quantized value using the quantization step size $\Delta$. Therefore, the embedding modulation is performed as follows:

$$d^* = (\lfloor \frac{d}{\Delta} \rfloor + (\lfloor \frac{d}{\Delta} \rfloor 2! = w)) \times \Delta \tag{3}$$

## 3.3  Watermark Embedding in WSS Data

In this work, we consider a framework which is composed by three entities: a Genomic Research Unity (GRU), a Genomic Research Center (GRC) and a Cloud Services Provider (CSP). GRU and GRC decide to outsource their genetic data on the cloud for storage and/or processing purposes. Before being outsourced, these data are watermarked so as to ensure their copyright protection and traitor tracing. To do so, we describe in this section a robust database watermarking scheme that allows message embedding for WSS data. Let us consider a WSS database $DB$, which consists of many genes, our solution is implemented as follow:

- First the table $DB$ is secretly reorganized into the database $DB^r$. To do so, data owner assigns a primary key $v_u.PK$ for each variant $v_u \in \{u = 1, \cdots, |variants|\}$, where $v_u.PK = CHROM\|POS\|GENE$. Then, this primary key is used for partitioning the database into $N_g$ groups using a secret watermarking key $K_w$. The group index number for each variant $n_{v_u}$ is computed based on secure hash algorithm using (4) and $N_g$ groups $\{G_i\}_{i=1,2,\cdots,N_g}$, are constituted.

$$n_{v_u} = H(K_w(H(v_u.PK|K_w)) \mod N_g \tag{4}$$

Once all groups are obtained, one bit of the watermark is embedded into each group.
- The data owner (in our case GRU/GRC) generates a binary watermark $W = \{w_1, w_2, \cdots, w_{N_g}\}$ uniformly distributed.
- Each group $G_i$ of the database is divided into two tuple sub-groups $G_i^A$ and $G_i^B$, based on the secret watermarking key $K_w$. To do so, the sub-group index number $n_{gv_u}$ for each variant $v_u$ in $G_i$, is computed using secure hash algorithm such that:

$$n_{gv_u} = H(K_w\|(H(v_u.PK\|K_w)) \mod 2 \tag{5}$$

If the value $n_{gv_u} = 1$, then the variant $v_u$ belongs to $G_i^A$, otherwise ($n_{gv_u} = 0$), then it belongs to $G_i^B$.
- QIM is used for embedding one watermark bit in these sub-groups so as to produce the watermarked sub-groups $G_{A,i}^W$ and $G_{B,i}^W$. The watermark embedding process is illustrated in Algorithm 1 according to three cases. After sub-group watermarking, the watermarked database $DB^{rw}$ is constituted.

## 3.4  Watermark Extraction

It is worth noting that during watermarking process, one watermark bit is embedded in each database column. Thus, during extraction stage a majority vote is performed in order to decide which watermark bit will be extracted. Indeed, majority vote is one

**Algorithm 1.** Watermark embedding modulation in one group
***
1: **INPUT**: Subgroups $G_i^A$ and $G_i^B$ of $G_i$, A watermark bit $w_i$, a quantization step size $\Delta$
2: **procedure** GROUPWATERMARKING($G_i^A, G_i^B, w_i, \Delta, d = |C_0^A| - |C_0^B|$)
3:    **if** $\lfloor \frac{d}{\Delta} \rfloor \% 2 == w_i$ **then**
4:        $d^* = \lfloor \frac{d}{\Delta} \rfloor \times \Delta$
5:    **else**
6:        $d^* = \lfloor \frac{d}{\Delta} \rfloor \times \Delta + \Delta$
7:    **end if**
8:    modulationValue = abs($d^* - \lfloor \frac{d}{\Delta} \rfloor$)
9: Case 1
10:    **if** $d^* \geq$ d and $|C_0^B| \geq$ modulationValue **then**
11:        $|C_0^{BW}| = |C_0^B|$ - modulationValue
12: Case 2
13:    **else if** $d^* <$ d and $|C_0^A| \geq$ modulationValue **then**
14:        $|C_0^{AW}| = |C_0^A|$ - modulationValue
15: Case 3
16:    **else**
17:        Not embeddable group
18:    **end if**
19:    **return** $G_{A,i}^W, G_{B,i}^W$
20: **end procedure**
***

of the popular optimal algorithms which is used to find the majority element among the given elements that have more than $\frac{N}{2}$ occurrences. However, watermark reading works similarly. The watermarked database $DB^w$ is first reorganized into $N_g$ groups, and each group $\{G_i^W\}_{i=1,\cdots,N_g}$ is partitioned into two sub-groups $G_{A,i}^W, G_{B,i}^W$. From each group, one message bit $w_{P_i}$ is detected and extracted in each column according to the Eq. (6). After that, a majority vote is conducted in order to decide which watermark bit is extracted. While tuple primary keys are not modified, the knowledge of the watermarking key ensures synchronization between watermark embedding and watermark detection/extraction. The watermark extraction process is illustrated in Algorithm 2. We discuss the theoretical performances of our solution in the next section before presenting experimental results.

$$w_{P_i} = \lfloor \frac{d^{w*} + \frac{\Delta}{2}}{\Delta} \rfloor \mod 2 \tag{6}$$

where

$$d^{w*} = |C_0^{AW}|_{P_i} - |C_0^{BW}|_{P_i}$$

## 4    Theoretical Performance

In this section, we start by presenting the constraints of some parameters in the proposed algorithm and then present the theoretical performance of our scheme in terms of distortion introduced to data during the watermarking, the insertion capacity, and the robustness against different database watermarking attacks.

**Algorithm 2.** Watermark extraction in one group

1: **INPUT**: Subgroups $G_{A,i}^W, G_{B,i}^W$ of $G_i^W$, a quantization step size $\Delta$
2: **procedure** WATERMARK DETECTION$(G_{A,i}^W, G_{B,i}^W, \Delta)$
3:     **for** each individual $(P_{i:i=1,\cdots,|patients|})$ in $G_i^W$ **do**
4:         $d^{w*} = |c_0^{AW}|_{Pi} - |c_0^{BW}|_{Pi}$
5:         $w_{Pi} = \lfloor \frac{d^{w*} + \frac{\Delta}{2}}{\Delta} \rfloor \bmod 2$
6:     **end for**
7:     $w_i = $ majority-vote$(w_{P_{i:i=1,\cdots,|patients|}})$
8:     **return** extracted watermark $(w_i)$
9: **end procedure**

## 4.1 Parameter Constraints

In our solution, in order to work properly and intuitively, some constraints such as distortion factor ($\Delta$), number of groups in the database ($N_g$), number of tuples in the database ($N_r$) and the probability to have 0 in one group ($Pr_0$) must to be defined and respected. These constraints are such that

$$\frac{S_g}{2} > \Delta \Leftrightarrow \frac{N_r}{N_g} > 2 \times \Delta \Leftrightarrow N_r > 2 \times \Delta \times N_g \Leftrightarrow Pr_0 \times N_r > 2 \times \Delta \times N_g \quad (7)$$

this constrain is important, because the number of zeros in a sub-group should be greater than the distortion factor $\Delta$ otherwise we can't embed the watermark into the group. As we will see later, this constraint will help us in analyzing the performance of our watermarking method.

## 4.2 Distortion Performance

Let us consider a database $DB$ which contains $N_c$ columns, $N_r$ rows and $db_{size}$ attribute values. During the watermarking process, this database is divided into $N_g$ groups, and each group is partitioned into two sub-groups. If $S_g$ is the number of tuples in one group. Then, the distortion value $D_\Delta$ for the database $DB$ corresponds to the number of modified attribute values in the database for a given $\Delta$, and can be computed as follows:

$$D_\Delta = N_g \times \frac{\Delta}{2} \times N_c \Longrightarrow D_\Delta = \frac{N_r}{S_g} \times \frac{\Delta}{2} \times N_c \Longrightarrow D_\Delta = db_{size} \times \frac{\Delta}{2 \times S_g} \quad (8)$$

As example, if we take $\Delta = 2$ and $S_g = 100$, then we can say that the distortion is $\frac{1}{100}$ of the $db_{size}$. This is due to symmetric distribution for the difference value of zero frequency between sub-group $G^A$ and sub-group $G^B$. This distortion does not disturb the results of WSS as we will see in Sect. 5.

## 4.3 Robustness Performance

In this section, we analyze the robustness of our watermarking scheme under two well-known database attacks that are deletion attacks and insertion attacks. We evaluate the robustness of our solution by means of the bit error rate (BER), which corresponds to the ratio of the number of incorrectly extracted watermark bits to the number of the original watermark bits. BER is such that:

$$BER = \frac{\sum_{i=1}^{N_g} w_i \oplus w_i'}{N_g} \qquad (9)$$

where $w_i$ and $w_i'$ are the embedded and the extracted watermark bit respectively. Lower value of BER means that we have a higher watermarking robustness. In the following, we discuss the attacks considered in this paper.

*Deletion Attacks :* Let us consider an attack that consists of a random deletion of attribute values or tuples in the database. We distinguish two cases for this attack: **(i) Column deletion:** in this case, an attacker tries to delete $N_{c_1}$ columns in the database. No matter how many columns are deleted, one column is enough to detect the watermark if all columns are watermarked. **(ii) Tuple deletion:** if the attacker randomly eliminates $N_d$ tuples in the database. The watermark may not be detected depending on the percentage of deleted data and the group to which deleted elements belongs. We will come back to this case in Sect. 5, where we demonstrated that the robustness of our solution against this attack using BER.

*Insertion Attacks:* An attacker may try to insert a certain number of columns or tuples in the database. Two cases differ. **(i) Column insertion:** An attacker tries to insert a certain number of columns in the database. By doing so, it requires an attacker to duplicate at least one time the number of columns (or individuals) so as to change the watermark bit. Assume that the original group verifies the probability to have 1 values is greater than the probability to have 0 values ($Pr_1 > Pr_0$). Then, the watermarked group verifies $Pr_0^w > Pr_1^w$. Hence, we can define $X = Pr_1 - Pr_0$ and $X^w = Pr_0^w - Pr_1^w$. There are three cases in which the data can be added by an attacker.

- **Case 1:** If $Pr_1 < Pr_0$, there is no problem as the attack will be always detected.
- **Case 2:** If $Pr_1 = Pr_0$, as in the previous case, the attack will always be detected.
- **Case 3:** If $Pr_1 > Pr_0$, the attacker requires to add $M$ elements in the database as:

$$M = \frac{N_c \times X^w}{X} \qquad (10)$$

**(ii) Tuple Insertion:** This attack corresponds to the insertion a certain number of tuples in the database. If $N$ is the number of tuples that the attacker want to insert in the database. Let $k$ be the number of success out of the total number of trials and $p$ the probability to succeed, while $q$ is the probability of failure. Thus, we have

$$p = \frac{1}{2 \times N_g} \quad , \quad q = 1 - p \qquad (11)$$

The probability of $k$ successes out of $N$ trials when the probability of one success is $p$ is computed according to the Eq. (12)

$$P(N, k, p) = \binom{N}{k} p^k q^{N-k} \qquad (12)$$

In the previous Eq. (12), the binomial coefficient express the number of combinations of $N$ takes $k$. It is calculated according to Eq. (13).

$$\binom{N}{k} = \frac{N!}{(N-k)!k!} \tag{13}$$

We give in next section, obtained results after simulating the discussed attacks.

# 5   Experimental Results and Discussion

We evaluate our watermarking method in terms of distortion, robustness, and insertion capacity in the case of a real genetic database.

## 5.1   Test Database

We used a genetic relational database composed of one table of 80 tuples issued from a real genetic database, from the FrEx project [29], that contains pieces of information related to genetic variants of 733 individuals. Such genetic variants are used by researchers or/and geneticists in GWAS [3] in order to determine if there is a relationship between these genetic variants and certain diseases. For each individual and each variant, the genotype corresponds to an integer value that takes the value 0 if the alternative allele is equal to the reference allele, 1 to the second alternative allele, and $k \in \{1, \cdots, g\}$ in case of $g$ possible alternative alleles. In the sequel, a set of attributes composed by the chromosome, the position, and the gene is considered as the primary key. We chose these attributes because their combination uniquely identifies each database tuple of variants.

## 5.2   Distortion Results

To test the impact of the proposed watermarking scheme for GWAS results, we have conducted a secure WSS method presented in [30]. In this context, the p-value has been used as a descriptive statistic, which is the p-value of the association, and the null hypothesis is that the allele frequencies at SNP are the same in cases and controls (For more details about the computation of the WSS p-value, please refer to [8]). To test our watermarking method, the database is divided into $N_g$ groups, considering several cases. These cases correspond to $N_g \in \{1, \cdots, 20\}$. We have also chosen different values of distortion step $\Delta$ such that $\Delta \in \{2, 4, 6, \cdots, 34\}$, each group is also divided into two sub-groups. In order to check if our experiments are going to be different in the results obtained from the experimental and control groups, Fig. 3 presents the p-values with different percentages of modulated data after applying our watermarking method on the given database. The obtained results in Fig. 3 set up such that they conveyed a meaning that there exists no distinction between the different samples and no interference in the association test results. Moreover, the results show us that the differences are real and not just due to chance as the p-value increases as the ratio of data distortion increases. In addition, the Table 3 presents the p-value results for above chosen $N_g$ and $\Delta$. The mentioned p-value are very low even with variable number of groups $N_g$ and quantization step $\Delta$. These results confirm their statistically significant and are less likely to be caused by noise.

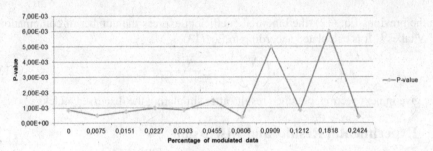

**Fig. 3.** Distortion percentage of modulated data.

**Table 3.** P-value results in function of the quantization step $\Delta$ and the number of groups $N_g$.

| $\frac{\Delta}{N_g}$ | 2 | 4 | 6 | 8 | 10 | 12 | 14 | 16 | 18 | 20 | 22 | 24 | 26 | 28 | 30 | 32 | 34 |
|---|---|---|---|---|---|---|---|---|---|---|---|---|---|---|---|---|---|
| 1 | $4.9 \times 10^{-4}$ | $3.5 \times 10^{-4}$ | $4.2 \times 10^{-4}$ | $4.4 \times 10^{-3}$ | $5.9 \times 10^{-4}$ | $2.9 \times 10^{-5}$ | $3.0 \times 10^{-4}$ | $5.9 \times 10^{-4}$ | $5.3 \times 10^{-4}$ | $5.4 \times 10^{-4}$ | $3.4 \times 10^{-3}$ | $7.9 \times 10^{-3}$ | $5.9 \times 10^{-4}$ | $3.9 \times 10^{-3}$ | $5.9 \times 10^{-3}$ | $3.9 \times 10^{-3}$ | $3.7 \times 10^{-4}$ |
| 2 | $5.9 \times 10^{-4}$ | $4.9 \times 10^{-4}$ | $6.6 \times 10^{-4}$ | $1.4 \times 10^{-3}$ | $4.2 \times 10^{-4}$ | $5.9 \times 10^{-3}$ | $5.9 \times 10^{-4}$ | $3.6 \times 10^{-4}$ | $9.9 \times 10^{-4}$ | $2.9 \times 10^{-3}$ | $3.8 \times 10^{-4}$ | $5.9 \times 10^{-3}$ | $2.9 \times 10^{-3}$ | $3.4 \times 10^{-3}$ | $2.3 \times 10^{-3}$ | $5.4 \times 10^{-3}$ | $9.9 \times 10^{-4}$ |
| 3 | $4.2 \times 10^{-4}$ | $1.9 \times 10^{-3}$ | $1.1 \times 10^{-3}$ | $4.9 \times 10^{-4}$ | $6.9 \times 10^{-4}$ | $1.1 \times 10^{-3}$ | $7.4 \times 10^{-4}$ | $1.1 \times 10^{-3}$ | $4.2 \times 10^{-4}$ | $5.8 \times 10^{-4}$ | $4.9 \times 10^{-4}$ | | | | | | |
| 4 | $6.6 \times 10^{-4}$ | $6.6 \times 10^{-4}$ | $3.9 \times 10^{-3}$ | $2.9 \times 10^{-3}$ | $5.2 \times 10^{-3}$ | $5.4 \times 10^{-4}$ | $5.9 \times 10^{-4}$ | $7.7 \times 10^{-4}$ | $2.4 \times 10^{-4}$ | | | | | | | | |
| 5 | $5.4 \times 10^{-4}$ | $7.4 \times 10^{-4}$ | $2.3 \times 10^{-4}$ | $8.5 \times 10^{-4}$ | $6.6 \times 10^{-4}$ | $2.9 \times 10^{-3}$ | $4.2 \times 10^{-4}$ | $8.7 \times 10^{-4}$ | | | | | | | | | |
| 6 | $5.9 \times 10^{-4}$ | $5.4 \times 10^{-4}$ | $9.9 \times 10^{-4}$ | $2.9 \times 10^{-3}$ | $2.9 \times 10^{-3}$ | $2.3 \times 10^{-3}$ | | | | | | | | | | | |
| 7 | $5.9 \times 10^{-4}$ | $6.6 \times 10^{-4}$ | $5.4 \times 10^{-4}$ | $1.3 \times 10^{-3}$ | $6.6 \times 10^{-4}$ | | | | | | | | | | | | |
| 8 | $5.4 \times 10^{-4}$ | $3.7 \times 10^{-4}$ | $4.6 \times 10^{-4}$ | $6.6 \times 10^{-4}$ | | | | | | | | | | | | | |
| 9 | $1.1 \times 10^{-3}$ | $4.4 \times 10^{-3}$ | $6.6 \times 10^{-4}$ | | | | | | | | | | | | | | |
| 10 | $3.3 \times 10^{-4}$ | $1.9 \times 10^{-3}$ | $2.3 \times 10^{-4}$ | | | | | | | | | | | | | | |
| 11 | $2.3 \times 10^{-3}$ | $3.5 \times 10^{-4}$ | | | | | | | | | | | | | | | |
| 12 | $7.7 \times 10^{-4}$ | $2.3 \times 10^{-3}$ | | | | | | | | | | | | | | | |
| 13 | $9.9 \times 10^{-4}$ | $4.2 \times 10^{-4}$ | | | | | | | | | | | | | | | |
| 14 | $2.6 \times 10^{-4}$ | $5.9 \times 10^{-4}$ | | | | | | | | | | | | | | | |
| 15 | $1.4 \times 10^{-3}$ | $8.9 \times 10^{-3}$ | | | | | | | | | | | | | | | |
| 16 | $7.4 \times 10^{-4}$ | | | | | | | | | | | | | | | | |
| 17 | $4.2 \times 10^{-4}$ | | | | | | | | | | | | | | | | |
| 18 | $1.9 \times 10^{-3}$ | | | | | | | | | | | | | | | | |
| 19 | $2.9 \times 10^{-3}$ | | | | | | | | | | | | | | | | |
| 20 | $4.2 \times 10^{-4}$ | | | | | | | | | | | | | | | | |

## 5.3   Capacity Results

The insertion capacity is evaluated by the ratio of database elements that can be used for the watermark embedding to the total number of elements in the database. Higher watermarking capacity means that more watermark information that we can embed in the database. The watermarking capacity of our solution depends on the number of embeddable groups that we have in the database. This capacity can reach 100% depending on genotypes that we have in the database. This means that in some cases, each group in the database can embed a watermark bit. However, if the capacity is the maximum, the robustness is reduced.

## 5.4   Robustness Results

We have simulated, different attacks on our watermarked database. We have considered an attacker that tries to insert, delete 10%, 20% and 30% of the data in the database. Obtained results are presented in Tables 4, 5, 6, 7, 8, 9. From these results, the watermark can be correctly detected from the database when BER approaches zero.

**Table 4.** BER results against column deletion 10%

| $\frac{c}{N_t}$ | 2 | 4 | 6 | 8 | 10 | 12 | 14 | 16 | 18 | 20 | 22 | 24 | 26 | 28 | 30 | 32 | 34 |
|---|---|---|---|---|---|---|---|---|---|---|---|---|---|---|---|---|---|
| 1 | 0 | 0 | 0 | 0 | 0 | 0 | 0 | 0 | 0 | 0 | 0 | 0 | 0 | 0 | 0 | 0 | 0 |
| 2 | 0 | 0 | 0 | 0 | 0 | 0 | 0 | 0 | 0 | 0 | 0 | 0.48 | 0.12 | 0.63 | 0.37 | 0.24 | |
| 3 | 0 | 0 | 0 | 0 | 0 | 0 | 0 | 0 | 0.47 | 0.27 | 0.25 | 0.43 | | | | | |
| 4 | 0 | 0 | 0 | 0 | 0 | 0.12 | 0.24 | 0.48 | 0.43 | | | | | | | | |
| 5 | 0 | 0 | 0 | 0 | 0.09 | 0.15 | 0.47 | 0.39 | | | | | | | | | |
| 6 | 0 | 0 | 0 | 0.09 | 0.23 | 0.31 | | | | | | | | | | | |
| 7 | 0 | 0 | 0.04 | 0.09 | 0.53 | | | | | | | | | | | | |
| 8 | 0 | 0 | 0.09 | 0.24 | | | | | | | | | | | | | |
| 9 | 0 | 0 | 0.16 | | | | | | | | | | | | | | |
| 10 | 0 | 0 | | | | | | | | | | | | | | | |
| 11 | 0 | 0 | | | | | | | | | | | | | | | |
| 12 | 0 | 0.02 | | | | | | | | | | | | | | | |
| 13 | 0 | 0.12 | | | | | | | | | | | | | | | |
| 14 | 0 | 0.17 | | | | | | | | | | | | | | | |
| 15 | 0 | | | | | | | | | | | | | | | | |
| 16 | 0 | | | | | | | | | | | | | | | | |
| 17 | 0 | | | | | | | | | | | | | | | | |
| 18 | 0.06 | | | | | | | | | | | | | | | | |
| 19 | 0.06 | | | | | | | | | | | | | | | | |
| 20 | 0.05 | | | | | | | | | | | | | | | | |

**Table 5.** BER results against column deletion 20%

| $\frac{c}{N_t}$ | 2 | 4 | 6 | 8 | 10 | 12 | 14 | 16 | 18 | 20 | 22 | 24 | 26 | 28 | 30 | 32 | 34 |
|---|---|---|---|---|---|---|---|---|---|---|---|---|---|---|---|---|---|
| 1 | 0 | 0 | 0 | 0 | 0 | 0 | 0 | 0 | 0 | 0 | 0 | 0 | 0 | 0 | 0 | 0 | 0 |
| 2 | 0 | 0 | 0 | 0 | 0 | 0 | 0 | 0 | 0 | 0 | 0 | 0.48 | 0.12 | 0.44 | 0.37 | 0.24 | |
| 3 | 0 | 0 | 0 | 0 | 0 | 0 | 0 | 0 | 0.47 | 0.27 | 0.25 | 0.43 | | | | | |
| 4 | 0 | 0 | 0 | 0 | 0 | 0.12 | 0.24 | 0.48 | 0.43 | | | | | | | | |
| 5 | 0 | 0 | 0 | 0 | 0.15 | 0.47 | 0.39 | | | | | | | | | | |
| 6 | 0 | 0 | 0 | 0.09 | 0.23 | 0.31 | | | | | | | | | | | |
| 7 | 0 | 0 | 0.04 | 0.09 | | | | | | | | | | | | | |
| 8 | 0 | 0 | 0.09 | | | | | | | | | | | | | | |
| 9 | 0 | 0 | | | | | | | | | | | | | | | |
| 10 | 0 | 0 | | | | | | | | | | | | | | | |
| 11 | 0 | 0 | | | | | | | | | | | | | | | |
| 12 | 0 | 0.05 | | | | | | | | | | | | | | | |
| 13 | 0 | 0.12 | | | | | | | | | | | | | | | |
| 14 | 0 | 0.15 | | | | | | | | | | | | | | | |
| 15 | 0.04 | | | | | | | | | | | | | | | | |
| 16 | 0 | | | | | | | | | | | | | | | | |
| 17 | 0 | | | | | | | | | | | | | | | | |
| 18 | 0.05 | | | | | | | | | | | | | | | | |
| 19 | 0.06 | | | | | | | | | | | | | | | | |
| 20 | 0.05 | | | | | | | | | | | | | | | | |

**Table 6.** BER results against column deletion 30%

| $\frac{c}{N_t}$ | 2 | 4 | 6 | 8 | 10 | 12 | 14 | 16 | 18 | 20 | 22 | 24 | 26 | 28 | 30 | 32 | 34 |
|---|---|---|---|---|---|---|---|---|---|---|---|---|---|---|---|---|---|
| 1 | 0 | 0 | 0 | 0 | 0 | 0 | 0 | 0 | 0 | 0 | 0 | 0 | 0 | 0 | 0 | 0 | 0 |
| 2 | 0 | 0 | 0 | 0 | 0 | 0 | 0 | 0 | 0 | 0 | 0 | 0.48 | 0.12 | 0.56 | 0.37 | 0.24 | |
| 3 | 0 | 0 | 0 | 0 | 0 | 0 | 0 | 0.15 | 0.47 | 0.27 | 0.25 | 0.43 | | | | | |
| 4 | 0 | 0 | 0 | 0 | 0 | 0.12 | 0.24 | 0.48 | 0.43 | | | | | | | | |
| 5 | 0 | 0 | 0 | 0 | 0.13 | 0.15 | 0.47 | | | | | | | | | | |
| 6 | 0 | 0 | 0 | 0.09 | 0.23 | 0.31 | | | | | | | | | | | |
| 7 | 0 | 0 | 0.04 | 0.09 | 0.53 | | | | | | | | | | | | |
| 8 | 0 | 0 | 0.09 | 0.24 | | | | | | | | | | | | | |
| 9 | 0 | 0 | 0.1 | | | | | | | | | | | | | | |
| 10 | 0 | 0 | | | | | | | | | | | | | | | |
| 11 | 0 | 0 | | | | | | | | | | | | | | | |
| 12 | 0 | 0.04 | | | | | | | | | | | | | | | |
| 13 | 0 | 0.12 | | | | | | | | | | | | | | | |
| 14 | 0 | 0.17 | | | | | | | | | | | | | | | |
| 15 | 0.04 | | | | | | | | | | | | | | | | |
| 16 | 0 | | | | | | | | | | | | | | | | |
| 17 | 0 | | | | | | | | | | | | | | | | |
| 18 | 0.02 | | | | | | | | | | | | | | | | |
| 19 | 0.06 | | | | | | | | | | | | | | | | |
| 20 | 0.05 | | | | | | | | | | | | | | | | |

**Table 7.** BER results against column addition 10%

| $\frac{c}{N_t}$ | 2 | 4 | 6 | 8 | 10 | 12 | 14 | 16 | 18 | 20 | 22 | 24 | 26 | 28 | 30 | 32 | 34 |
|---|---|---|---|---|---|---|---|---|---|---|---|---|---|---|---|---|---|
| 1 | 0 | 0 | 0 | 0 | 0 | 0 | 0 | 0 | 0 | 0 | 0 | 0 | 0 | 0 | 0 | 0 | 0 |
| 2 | 0 | 0 | 0 | 0 | 0 | 0 | 0 | 0 | 0 | 0 | 0 | 0.48 | 0.12 | 0.69 | 0.37 | 0.24 | |
| 3 | 0 | 0 | 0 | 0 | 0 | 0 | 0 | 0.19 | 0.47 | 0.27 | 0.25 | 0.43 | | | | | |
| 4 | 0 | 0 | 0 | 0 | 0 | 0.12 | 0.36 | 0.48 | 0.43 | | | | | | | | |
| 5 | 0 | 0 | 0 | 0 | 0.14 | 0.15 | 0.47 | | | | | | | | | | |
| 6 | 0 | 0 | 0.04 | 0.09 | 0.23 | 0.31 | | | | | | | | | | | |
| 7 | 0 | 0 | 0.09 | 0.24 | 0.53 | | | | | | | | | | | | |
| 8 | 0 | 0 | 0.19 | 0.24 | | | | | | | | | | | | | |
| 9 | 0 | 0 | 0.20 | | | | | | | | | | | | | | |
| 10 | 0 | 0 | | | | | | | | | | | | | | | |
| 11 | 0 | 0 | | | | | | | | | | | | | | | |
| 12 | 0 | 0.10 | | | | | | | | | | | | | | | |
| 13 | 0 | 0.12 | | | | | | | | | | | | | | | |
| 14 | 0 | 0.17 | | | | | | | | | | | | | | | |
| 15 | 0.04 | | | | | | | | | | | | | | | | |
| 16 | 0 | | | | | | | | | | | | | | | | |
| 17 | 0 | | | | | | | | | | | | | | | | |
| 18 | 0.06 | | | | | | | | | | | | | | | | |
| 19 | 0.06 | | | | | | | | | | | | | | | | |
| 20 | 0.05 | | | | | | | | | | | | | | | | |

**Table 8.** BER results against column addition 20%

| $\frac{c}{N_t}$ | 2 | 4 | 6 | 8 | 10 | 12 | 14 | 16 | 18 | 20 | 22 | 24 | 26 | 28 | 30 | 32 | 34 |
|---|---|---|---|---|---|---|---|---|---|---|---|---|---|---|---|---|---|
| 1 | 0 | 0 | 0 | 0 | 0 | 0 | 0 | 0 | 0 | 0 | 0 | 0 | 0 | 0 | 0 | 0 | 0 |
| 2 | 0 | 0 | 0 | 0 | 0 | 0 | 0 | 0 | 0 | 0 | 0 | 0.40 | 0.48 | 0.12 | 0.69 | 0.37 | 0.24 |
| 3 | 0 | 0 | 0 | 0 | 0 | 0 | 0 | 0 | 0.51 | 0.47 | 0.27 | 0.45 | 0.43 | | | | |
| 4 | 0 | 0 | 0 | 0 | 0 | 0.12 | 0.36 | 0.48 | 0.43 | | | | | | | | |
| 5 | 0 | 0 | 0 | 0 | 0.05 | 0.14 | 0.24 | 0.47 | 0.39 | | | | | | | | |
| 6 | 0 | 0 | 0 | 0.09 | 0.24 | | | | | | | | | | | | |
| 7 | 0 | 0 | 0.04 | 0.24 | 0.53 | | | | | | | | | | | | |
| 8 | 0 | 0 | 0.19 | 0.24 | | | | | | | | | | | | | |
| 9 | 0 | 0 | 0.25 | | | | | | | | | | | | | | |
| 10 | 0 | 0 | | | | | | | | | | | | | | | |
| 11 | 0 | 0 | | | | | | | | | | | | | | | |
| 12 | 0 | 0.12 | | | | | | | | | | | | | | | |
| 13 | 0 | 0.14 | | | | | | | | | | | | | | | |
| 14 | 0 | 0.17 | | | | | | | | | | | | | | | |
| 15 | 0 | | | | | | | | | | | | | | | | |
| 16 | 0 | | | | | | | | | | | | | | | | |
| 17 | 0 | | | | | | | | | | | | | | | | |
| 18 | 0.06 | | | | | | | | | | | | | | | | |
| 19 | 0.06 | | | | | | | | | | | | | | | | |
| 20 | 0.05 | | | | | | | | | | | | | | | | |

**Table 9.** BER results against column addition 30%

| $\frac{c}{N_t}$ | 2 | 4 | 6 | 8 | 10 | 12 | 14 | 16 | 18 | 20 | 22 | 24 | 26 | 28 | 30 | 32 | 34 |
|---|---|---|---|---|---|---|---|---|---|---|---|---|---|---|---|---|---|
| 1 | 0 | 0 | 0 | 0 | 0 | 0 | 0 | 0 | 0 | 0 | 0 | 0 | 0 | 0 | 0 | 0 | 0 |
| 2 | 0 | 0 | 0 | 0 | 0 | 0 | 0 | 0 | 0 | 0 | 0 | 0.40 | 0.48 | 0.55 | 0.69 | 0.37 | 0.24 |
| 3 | 0 | 0 | 0 | 0 | 0 | 0 | 0 | 0 | 0.11 | 0.51 | 0.47 | 0.27 | 0.45 | 0.43 | | | |
| 4 | 0 | 0 | 0 | 0 | 0 | 0.11 | 0.37 | 0.50 | 0.64 | | | | | | | | |
| 5 | 0 | 0 | 0 | 0 | 0.05 | 0.35 | 0.25 | 0.74 | 0.39 | | | | | | | | |
| 6 | 0 | 0 | 0 | 0.09 | 0.24 | | | | | | | | | | | | |
| 7 | 0 | 0 | 0.04 | | | | | | | | | | | | | | |
| 8 | 0 | 0 | 0.19 | | | | | | | | | | | | | | |
| 9 | 0 | 0 | 0.22 | | | | | | | | | | | | | | |
| 10 | 0 | 0 | | | | | | | | | | | | | | | |
| 11 | 0 | 0 | | | | | | | | | | | | | | | |
| 12 | 0 | 0.12 | | | | | | | | | | | | | | | |
| 13 | 0 | 0.14 | | | | | | | | | | | | | | | |
| 14 | 0 | | | | | | | | | | | | | | | | |
| 15 | 0 | | | | | | | | | | | | | | | | |
| 16 | 0 | | | | | | | | | | | | | | | | |
| 17 | 0 | | | | | | | | | | | | | | | | |
| 18 | 0.07 | | | | | | | | | | | | | | | | |
| 19 | 0.07 | | | | | | | | | | | | | | | | |
| 20 | 0.08 | | | | | | | | | | | | | | | | |

## 6 Conclusion

In this paper, we have proposed a new robust database watermarking method that allows watermarking of genomic data used in GWAS. It is the first method of this kind, and it can be used for statistical algorithms such as the WSS method. It can be used in protecting traitor tracing and copyright protection, and it is based on QIM and majority vote. We have studied theoretical performance and experimentally verified the performance of our solution in terms of robustness against deletion and addition attacks, insertion capacity, and distortion. In this method, a watermark is embedded in genetic data without altering the results of association tests that can be conducted on these data. This comfort its future use in real-life applications, especially in cloud environments. As the primary form of the proposed modulation technique is to preserving the statistical analysis of GWA studies, we plan to study the proposed technique for other GWA studies in the future.

## References

1. Mehrgou, A., Akouchekian, M.: The importance of BRCA1 and BRCA2 genes mutations in breast cancer development. Med. J. Islamic Repub. Iran (MJIRI) **30**(369), 1–12 (2016)
2. Ginsburg, G.: Medical genomics: gather and use genetic data in health care. Nat. News **508**(7497), 451–453 (2014)
3. Wang, M.H., Cordell, H.J., Van Steen, K.:Statistical methods for genome-wide association studies. In: Seminars in Cancer Biology, vol. 55, pp. 53–60. Elsevier (2019)
4. Taleb, A., Kirchler, M., Monti, R., Lippert, C.: ContiG: self-supervised multimodal contrastive learning for medical imaging with genetics. In: Proceedings of the IEEE/CVF Conference on Computer Vision and Pattern Recognition (CVPR), pp. 20 908–20 921. IEEE (2022)
5. Michael, B.E., Yann, L.G., Sarah, E.J., Napolioni, V., Michael, G.D., Zihuai, H.: A fast and robust strategy to remove variant-level artifacts in alzheimer disease sequencing project data. Neurol. Genet. **8**(5), e200012 (2022)
6. Shin, J., et al.: PhenGenVar: a user-friendly genetic variant detection and visualization tool for precision medicine. J. Personalized Med. **12**(6), 1–11 (2022)
7. Ozaki, K., et al.: Functional SNPs in the lymphotoxin-$\alpha$ gene that are associated with susceptibility to myocardial infarction. Nat. Genet. **32**(4), 650–654 (2002)
8. Madsen, B.E., Browning, S.R.: A groupwise association test for rare mutations using a weighted sum statistic. PLoS Genet. **5**(2), 1–11 (2009)
9. Ding, H., Tian, Y., Peng, C., Zhang, Y., Xiang, S.: Inference attacks on genomic privacy with an improved HMM and an RCNN model for unrelated individuals. Inf. Sci. **512**, 207–218 (2020)
10. Bellafqira, R., Coatrieux, G., Genin, E., Cozic, M.: Secure multilayer perceptron based on homomorphic encryption. In: Yoo, C.D., Shi, Y.-Q., Kim, H.J., Piva, A., Kim, G. (eds.) IWDW 2018. LNCS, vol. 11378, pp. 322–336. Springer, Cham (2019). https://doi.org/10.1007/978-3-030-11389-6_24
11. Rady, M., Abdelkader, T., Ismail, R.: Integrity and confidentiality in cloud outsourced data. Ain Shams Eng. J. **10**, 275–285 (2019)
12. Wang, X., Jiang, X., Vaidya, J.: Efficient verification for outsourced genome-wide association studies. J. Biomed. Inform. **117**, 103714 (2021)

13. Wang, J., Du, X., Lu, J., Lu, W.: Bucket-based authentication for outsourced databases. Concurrency Comput. Pract. Experience **22**(9), 1160–1180 (2010)
14. Niyitegeka, David, Coatrieux, Gouenou, Bellafqira, Reda, Genin, Emmanuelle, Franco-Contreras, Javier: Dynamic watermarking-based integrity protection of homomorphically encrypted databases – application to outsourced genetic data. In: Yoo, Chang D.., Shi, Yun-Qing., Kim, Hyoung Joong, Piva, Alessandro, Kim, Gwangsu (eds.) IWDW 2018. LNCS, vol. 11378, pp. 151–166. Springer, Cham (2019). https://doi.org/10.1007/978-3-030-11389-6_12
15. Boujdad, F.-Z., Niyitegeka, D., Bellafqira, R., Coatrieux, G., Génin, E., Südholt, M.S.: A hybrid cloud deployment architecture for privacy-preserving collaborative genome-wide association studies. In: Gladyshev, P., Goel, S., James, J., Markowsky, G., Johnson, D. (eds.) ICDFC 2021. LNICST, vol. 441, pp. 342–359. Springer, Cham (2022). https://doi.org/10.1007/978-3-031-06365-7_21
16. Chen, W.: An artificial chromosome for data storage. Nat. Sci. Rev. **8**(5), nwab028 (2021)
17. Nguyen, T.T., Cai, K., Song, W., Immink, K.A.S.: Optimal single chromosome-inversion correcting codes for data storage in live DNA. In: IEEE International Symposium on Information Theory (ISIT), pp. 1791–1796. IEEE (2022)
18. Vinodhini, R., Malathi, P.: Hiding information in the DNA sequence using DNA steganographic algorithms with double-layered security. Int. J. Inf. Secur. Priv. (IJISP) **16**(1), 1–20 (2022)
19. Wang, Y., Han, Q., Cui, G., Sun, J.: Hiding messages based on DNA sequence and recombinant DNA technique. IEEE Trans. Nanotechnol. **18**, 299–307 (2019)
20. Lee, S.-H.: Reversible data hiding for DNA sequence using multilevel histogram shifting. Secur. Commun. Netw. **2018**, 1–13 (2018)
21. Hamad, S., Elhadad, A., Khalifa, A.: DNA watermarking using codon postfix technique. IEEE/ACM Trans. Comput. Biol. Bioinf. **15**(5), 1605–1610 (2017)
22. Ayday, E., Yilmaz, E., Yilmaz, A.: Robust optimization-based watermarking scheme for sequential data. In: $22^{nd}$ International Symposium on Research in Attacks, Intrusions and Defenses, pp. 323–336 (2019)
23. Kuribayashi, M., Fukushima, T., Funabiki, N.: Robust and secure data hiding for PDF text document. IEICE Trans. Inf. Syst. **102**(1), 41–47 (2019)
24. Pabinger, S., et al.: A survey of tools for variant analysis of next-generation genome sequencing data. Brief. Bioinform. **15**(2), 256–278 (2014)
25. Danecek, P.: The variant call format and VCF tools. Bioinformatics **27**(15), 2156–2158 (2011)
26. Rani, S., Halder, R.: Comparative analysis of relational database watermarking techniques: an empirical study. IEEE Access **10**, 27970–27989 (2022)
27. Li, Y., Guo, H., Jajodia, S.: Tamper detection and localization for categorical data using fragile watermarks. In: Proceedings of the $4^{th}$ ACM Workshop on Digital Rights Management, pp. 73–82 (2004)
28. Chen, B., Wornell, G.W.: Quantization index modulation: a class of provably good methods for digital watermarking and information embedding. IEEE Trans. Inf. Theory **47**(4), 1423–1443 (2001)
29. Genin, E., Redon, R., Deleuze, J.-F., Campion, D., Lambert, J.-C., Dartigues, J.-F.: The French exome (FREX) project: a population-based panel of exomes to help filter out common local variants. Int. Genet. Epidemiol. Soc. **41**, 691 (2017)
30. Bellafqira, R., Ludwig, T.E., Niyitegeka, D., Génin, E., Coatrieux, G.: Privacy-preserving genome-wide association study for rare mutations-a secure framework for externalized statistical analysis. IEEE Access **8**, 112515–112529 (2020)

# Adaptive Robust Watermarking Method Based on Deep Neural Networks

Fan Li[1,3], Chen Wan[1,3], and Fangjun Huang[2,3(✉)]

[1] School of Computer Science and Engineering, Sun Yat-sen University, Guangzhou 510006, China

[2] School of Cyber Science and Technology, Sun Yat-sen University, Shenzhen 518107, China
huangfj@mail.sysu.edu.cn

[3] Guangdong Provincial Key Laboratory of Information Security Technology, Sun Yat-sen University, Guangzhou 510006, China

**Abstract.** Aiming at the problem of digital multimedia piracy and infringement, an adaptive robust watermarking algorithm based on Deep Neural Networks (DNNs) is proposed. In our method, the watermark sequence to be embedded is mapped to a noise pattern first, which has the same dimension as the carrier image. Specifically, the noise pattern is generated adaptively according to the statistical properties of the carrier image, in which the noise intensity corresponding to the texture area of the carrier image is large, and that corresponding to the smooth area is small. Thus, after adding the generated noise pattern to the carrier image, good visual quality can be easily obtained. Furthermore, considering a series of attacks such as adding noise and JPEG compression, the watermark encoder and decoder in our scheme are jointly trained to resist the potential attacks in the physical world. Experimental results demonstrate that better visual quality and higher robustness can be obtained compared with those state-of-the-art algorithms based on DNNs. This means that we have better solved the problem of mutual restriction between visual quality and robustness.

**Keywords:** Robust watermarking · Adaptive strategy · Deep neural networks

## 1 Introduction

Nowadays, image is one of the most popular carriers of information exchange. The copyright of digital images has attracted more and more attention, and how to protect the copyright of digital images has become an urgent problem to be solved [1].

In the past few years, with the great success of deep learning in the fields of computer vision and pattern recognition, a series of digital steganography [2–7] and watermarking methods [7–16] based on deep neural networks (DNNs) have been proposed. In this paper, we mainly focus on robust watermarking technology. It can embed the watermark information into the carrier image in a visually imperceptible way. However, when the image is disturbed by some image processing attacks such as adding noise and scaling, the watermark information can be extracted reliably. The two key points to measure the performance of a watermarking system are robustness and visual imperceptibility.

X. Zhao et al. (Eds.): IWDW 2022, LNCS 13825, pp. 162–173, 2023.
https://doi.org/10.1007/978-3-031-25115-3_11

HiDDeN [7] is the first end-to-end trainable framework to realize robust watermarking using DNNs. In this framework, the encoder network receives a cover image and a watermark sequence, and outputs an encoded image; the decoder network receives the encoded image and attempts to reconstruct the watermark message. Considering HiD-DeN's lack of robustness to some specific attacks in the physical world, StegaStamp [11] adopt the image perturbation module in the training process, which approximately simulates the display and imaging process of the real world. In [12], by introducing a dataset of 1,000,000 camera captured images, Wengrowski et al. designed a camera display transfer function (CDTF), which can be used to model the camera-display pipeline for training the encoder and decoder in their method. Luo et al. [13] pointed out that with resorting to the neural network channel coding, the decoding accuracy of watermark message can be improved for those deep learning-based watermarking schemes. In [15], Zhang et al. proposed the universal deep hiding meta-architecture (UDH), in which the watermark message generated by their method is independent of the carrier image. In [16], Abdelnabi et al. first proposed an end-to-end model to hide data in the text, which is effective in largely preserving text utility and decoding the watermark while hiding its presence against adversaries.

In the field of robust watermarking, visual quality, robustness and embedding capacity are constrained by each other. Existing methods combine visual quality and robustness loss for training, but they do not limit the embedding position, and the comprehensive performance of visual quality and robustness is poor. There is an urgent need for a method to achieve better visual quality and robustness at the same time.

In this paper, we propose a new adaptive robust watermarking algorithm based on DNNs. In our method, the watermark sequence to be embedded is mapped to a noise pattern first, which is generated adaptively according to the statistical properties of the carrier image. Specifically, the noise intensity corresponding to the texture area of the carrier image is large, and that corresponding to the smooth area is small. Thus, after adding the generated noise pattern to the carrier image, good visual quality can be easily obtained.

## 2 Proposed Method

The proposed framework consists of two modules, the watermark embedding module and the watermark decoding module, which are depicted in Fig. 1. The embedding module is contained in the green dotted line box and the decoding module is in the red dotted line box. Although the embedding and decoding modules are independent, they are trained together during the training process. Following that, the proposed method will be thoroughly described from three perspectives: watermark embedding, watermark decoding, and loss function.

### 2.1 Watermark Embedding Module

The purpose of the watermark embedding module is to generate the watermarked image with high visual quality, which mainly includes two components: encoder E and threshold generation part TGP.

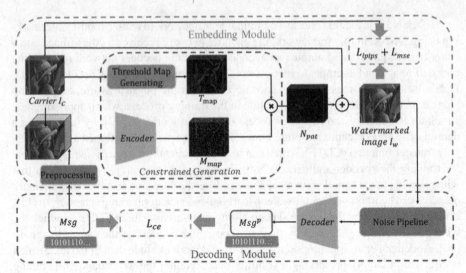

**Fig. 1.** Our robust watermarking framework.

The encoder $E$ is implemented with the Unet++ [17], and the output of the last layer adopts the sigmoid function, which can normalize the output values to [0, 1]. The encoder $E$ receives a carrier image $I_c$ of shape H × W × C and the preprocessed watermark information, and outputs the modification map $M_{map}$, which has the same shape as the input carrier image $I_c$. Note that the binary watermark sequence $Msg \in \{0, 1\}^L$ of length $L$ is not input into the encoder E directly. It should be preprocessed through a fully connected network and then up-sampled to produce a H × W × C tensor as in [11].

The Threshold Generation Part (TGP) receives the carrier image $I_c$ as input and produces the distortion threshold map $T_{map}$. Specifically, we perform Gaussian filtering on the carrier image $I_c$ to obtain image $I_g$ first. Then, the Sobel operators are used to calculate the first-order differences of $I_g$ in horizontal and vertical directions to obtain the threshold map $T_{map}$ as shown in Eq. (1).

$$T_{map} = \left|I_g \otimes K_h\right| + \left|I_g \otimes K_v\right| \tag{1}$$

where $\otimes$ represents convolution operation, and $K_h$ and $K_v$ are Sobel operators in horizontal and vertical directions, respectively.

After obtaining $M_{map}$ and $T_{map}$, the specific noise pattern $N_{pat}$ can be generated by performing the element-wise product operation on $M_{map}$ and $T_{map}$ as shown in Eq. (2). With adding $N_{pat}$ to the carrier image $I_c$, we can obtain the watermarked image $I_w$, as shown in Eq. (3).

$$N_{pat} = M_{map} \cdot T_{map} \tag{2}$$

$$I_w = I_c + \varepsilon * N_{pat} \tag{3}$$

where $T_{map} \in \mathbb{R}^{H \times W \times C}, M_{map} \in \mathbb{R}^{H \times W \times C}$, and $\varepsilon$ is a parameter that controls the embedding strength of $N_{pat}$.

## 2.2 Watermark Decoding Module

The watermark decoding module includes Noise PipeLine (NPL) and decoder D. The NPL simulates the noise in the real physical world by performing a series of attacks on the obtained watermarked image $I_w$. The input of NPL is the watermarked image $I_w$, and the output is the attacked image $I_w^*$. The decoder D consists of 7 convolution layers and 2 full connection layers, which receives the attacked image $I_w^*$ as input and outputs the predicted watermark sequence $Msg^p$.

It should be emphasized that NPL attacks the watermarked image $I_w$ by simulating various distortions in the real physical world to improve the robustness of the decoder. Since the physical attacks are composed of multiple single attacks, we decompose them into single attacks, which are analyzed separately. In our scheme, the following single attacks are included.

**Moire Fringe Noise**
In order to resist Moire fringe noise, we generate Moire noise M and add it to watermarked image to simulate Moire noise attack, as shown in Eq. (4).

$$I_w^* = (1 - \alpha)I_w + \alpha M \tag{4}$$

In Eq. (4), $\alpha \in [0, 1]$ refers to the transparency of the foreground area relative to the background image [18], and $I_w$ and $I_w^*$ represent the watermarked image and attacked image, respectively.

**Perspective Warp**
Using the Perspective warp, we can simulate the geometric transformation attack during photographing. The new coordinates $(x', y')$ is modified within a fixed range, as shown Eq. (5).

$$|x - x'| < W * \tau, |y - y'| < H * \tau \tag{5}$$

where $(x, y)$ represents the original coordinates, $(x', y')$ represents the new coordinates, $W$ and $H$ are the width and height of the image, $\tau$ represents the scale of the allowable offset of the coordinates.

**JPEG Compression**
Let $S$ represents the $8 \times 8$ Discrete Cosine Transform (DCT) block in JPEG (Joint Photographic Experts Group) compression process and $S_{i,j}(0 \leq i \leq 7, 0 \leq j \leq 7)$ represents the DCT coefficient in $S$. We simulate the quantization and rounding operations in JPEG compression process, as shown in Eq. (6).

$$B_{i,j} = f\left(\frac{S_{i,j}}{\beta Q_{i,j}}\right) \tag{6}$$

where $Q_{i,j}(0 \leq i \leq 7, 0 \leq j \leq 7)$ represents the element in the standard JPEG quantization table with the quality factor (QF) of 50, and $\beta$ is a parameter used to control the quantitative loss. The larger $\beta$ means the higher compression rate and the stronger the JPEG compression attack, which also leads to the worse image quality. The function $f$ is defined as $f(x) = x^3, x < 0.5; f(x) = x, x > 0.5$, which helps to reduce the phenomenon of gradient disappearance.

**Other Common Noise Attacks**

In addition to the above-mentioned attacks, we use a series of random color transformations to approximate the display-shooting perturbations, e.g., Gaussian noise $N(\mu, \sigma^2)$, Brightness and Contrast offset $bI_w + c$, etc. Note that after these attacks, the attacked $I_w^*$ should be clipped to $[0, 1]$.

## 2.3 Loss Function

In order to make sure that the watermarked image $I_w$ looks visually similar to carrier image $I_c$, two losses, i.e., Learned Perceptual Image Patch Similarity (LPIPS) loss $L_{lpips}$ [19] and Mean Square Error (MSE) loss $L_{mse}$, are employed to measure the "similarity" between $I_c$ and $I_w$.

In addition, considering that the decoded watermark $Msg^p$ should be the same as the original embedded watermark $Msg$, the difference between $Msg$ and $Msg^p$ is evaluated by the cross-entropy loss $L_{ce}$. The final training loss $L_{total}$ in our scheme is shown in Eq. (7).

$$L_{total} = \lambda_1 L_{ce} + \lambda_2 L_{mse} + \lambda_3 L_{lpips} \tag{7}$$

where $\lambda_1$, $\lambda_2$, and $\lambda_3$ represent the weights of $L_{ce}$, $L_{mse}$, and $L_{lpips}$, respectively.

# 3 Experiments

In our experiments, 25,000 images randomly selected from Mirflickr [20] are used as the training data, 1500 images randomly selected from the ImageNet2012 [21] validation set are used for testing, and all images are scaled to the size of $400 \times 400 \times 3$. The watermark encoder and decoder are jointly trained with the training data first, and then the obtained model is evaluated with those images in the test dataset.

## 3.1 Training Process

In the training process, the hyper-parameters for different attacks are shown in Table 1, where $p \sim U$ means that the parameter $p$ is a number randomly selected from U. In order to accelerate the optimization speed, the Adam optimizer [22] is utilized in our training process, where the learning rate and the number of epochs are set as 0.00001 and 200 respectively. Some details that may promote the convergence of the network are summarized below.

- To ensure that the embedded watermark sequence can be accurately decoded, only $L_{ce}$ loss is optimized at the beginning of training process, i.e., in the first 8000 steps of training, This can quickly reduce the overall loss and prevent the gradient from disappearing and unable to train. The training loss is set as $L_{total} = \lambda_1 L_{ce}$ with $\lambda_1 = 3.5$. Then in the next training process, the training loss is set as $L_{total} = \lambda_1 L_{ce} + \lambda_2 L_{mse} + \lambda_3 L_{lpips}$ with $\lambda_1 = 3.5$, $\lambda_2 = 2$, and $\lambda_3 = 1.5$.

- The adaptive embedding parameter $\varepsilon$ needs to be set to a large value in the early stage and gradually decreases in the later stage. In this paper, we set $\varepsilon = 0.8$ in the first 5 epochs, and then gradually decays to 0.25 linearly.

**Table 1.** The parameters for various attacks.

| Attacks | Parameters |
|---------|-----------|
| Moire fringe noise | $\alpha = 0.3$ |
| Perspective warp | $\tau = 0.1$ |
| JPEG compression | $\beta \sim U\ [0.1, 1.1]$ |
| Gaussian noise | $\mu = 0, \sigma \sim U\ [0, 0.18]$ |
| Brightness and Contrast | $b \sim U\ [0.6, 1.4], c \sim U\ [-0.2, 0.2]$ |

### 3.2  Test Results

To demonstrate the efficiency of our new method, Peak Signal to Noise Ratio (PSNR) [23], Structural Similarity (SSIM) [24], and Learned Perceptual Image Patch Similarity (LPIPS) [19] are selected in the test process to evaluate the visual quality of watermarked images. Note that for PSNR and SSIM, higher is better, and for LPIPS, lower is better. In addition, we test the bit recovery accuracy of the watermark sequence under various attacks. The bit recovery accuracy (ACC) is equal to $n/L$, where $n$ is the number of bits correctly recovered by the decoder, and $L$ represents the total length of the watermark sequence. Two state-of-the-art methods, HiDDeN [7] and StegaStamp [11] are selected for comparison in our experiments.

**Visual Quality**
The binary watermark sequences used in our experiments are randomly generated with the length of 50, 100, 150, and 200 bits respectively. The PSNR, SSIM, and LPIPS values between the carrier image $I_c$ and the watermarked image $I_w$ are shown in Table 2, and some watermarked images after embedding 100 bits of watermark information are shown in Fig. 2.

As shown in Table 2, the visual quality of watermarked images generated by our method is better than that generated by HiDDeN [7] and StegaStamp [11] when the embedded message is less than 100 bits. However, when the embedded message is more than 100 bits, the visual quality of watermarked images generated by our method is slightly worse than HiDDeN [7] but still better than StegaStamp [11]. It is generally known that embedding capacity, visual quality and robustness are mutually exclusive. Therefore, when the embedding capacity is increased, the performance of visual quality and robustness will be reduced accordingly. However, the comprehensive performance of this paper is better than the existing methods.

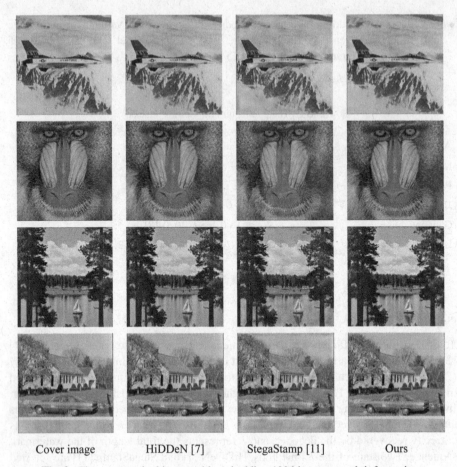

Cover image　　HiDDeN [7]　　StegaStamp [11]　　Ours

**Fig. 2.** The watermarked image with embedding 100 bits watermark information.

**Table 2.** The PSNR, SSIM and LPIPS values between carrier image and watermarked image.

| L | Method | PSNR↑ | SSIM↑ | LPIPS↓ |
|---|--------|-------|-------|--------|
| 50 | HiDDeN [7] | 34.52 | 0.9494 | 0.0275 |
| | StegaStamp [11] | 30.87 | 0.9258 | 0.0697 |
| | Ours | **36.46** | **0.9756** | **0.0097** |
| 100 | HiDDeN [7] | 36.95 | 0.9745 | 0.0102 |
| | StegaStamp [11] | 28.59 | 0.8949 | 0.0885 |
| | Ours | **37.46** | **0.9781** | **0.0078** |
| 150 | HiDDeN [7] | **39.89** | **0.9884** | **0.0039** |

(*continued*)

**Table 2.** (*continued*)

| L | Method | PSNR↑ | SSIM↑ | LPIPS↓ |
|---|---|---|---|---|
| | StegaStamp [11] | 30.66 | 0.9260 | 0.0683 |
| | Ours | 33.71 | 0.9461 | 0.0189 |
| 200 | HiDDeN [7] | **40.03** | **0.9963** | **0.0017** |
| | StegaStamp [11] | 31.64 | 0.9409 | 0.0631 |
| | Ours | 33.30 | 0.9339 | 0.0199 |

In Fig. 3, we also illustrated the specific noise pattern $N_{pat}$ that has been generated in the embedding process, which is added to the carrier image to obtain the watermarked image. Note that the noise pattern $N_{pat}$ is obtained by performing the dot product operation on the modification map $M_{map}$ and threshold map $T_{map}$ as shown in Eq. (2). In general, the elements in $N_{pat}$ are with small values, which are magnified 10 times for visualization in Fig. 3. As seen, when using our method to embed the watermark, the modification mainly focuses on the edge or the texture area of the carrier image (e.g., the hair of Lena images, the tree part in the house image, and the blue snow mountain of Aircraft image), which may result in the better visual quality of the watermarked image.

**Bit Recovery Accuracy**

In order to demonstrate the robustness of our method, a series of attacks are conducted on the watermarked image $I_w$ separately first, and then we combine all the attacks together attacked the watermarked image $I_w$, which is also called combined attack. The bit recovery accuracies under different attacks are shown in Table 3, where B&C represents Brightness and Contrast offset ($b\sim U$ [0.5, 1.5], $c\sim U$ [0.1, 0.3]), Gauss represents Gaussian noise attack ($\mu = 0$, $\sigma\sim U$ [0, 0.2]), JPEG represents JPEG compression ($\beta\sim U$ [0.1, 1.1]), Moire represents Moire fringe noise attack ($\alpha\sim U$ [0, 0.3]), Warp represents Perspective warp attack ($\tau = 0.1$), and Combined represents the combined attack with all the above parameters. The experimental results demonstrate that under different attacks, the bit recovery accuracy rates of our method are higher than that of StegaStamp and HiDDeN. Especially with the increase of the embedding rate, our method shows stronger robustness.

**The Influence of Attack Intensity on Bit Recovery Accuracy**

In order to explore the impact of the attack intensity on watermark recovery accuracy, we also carry out the following experiments, where the binary watermark sequence to be embedded is with the length of 100.

The experimental results are shown in Fig. 4, where the abscissa represents the attack intensity, and the ordinate represents the bit recovery accuracy. The experimental results corresponding to JPEG compression, Perspective warp, Moire fringe noise, Gaussian noise, and Brightness and Contrast are shown in Fig. 4(a)–(e), respectively. As seen, with the increase of attack intensity, the bit recovery accuracy will decrease. The Perspective wrap has the greatest impact on the bit recovery rate, while Moire fringe noise has small impact on the bit recovery rate. In addition, we can find that our proposed method and

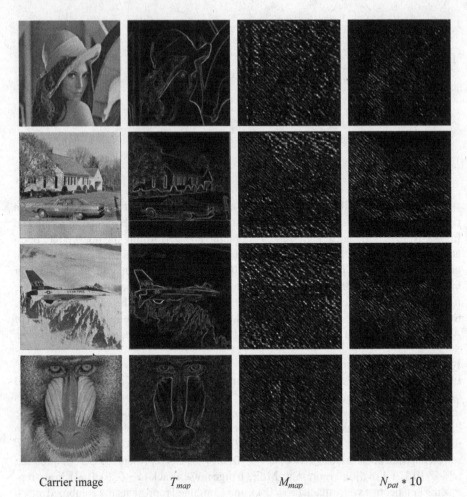

Carrier image          $T_{map}$          $M_{map}$          $N_{pat} * 10$

**Fig. 3.** Visualization results of the embedding process. (Color figure online)

**Table 3.** Bit recovery accuracy under different attacks.

| L | Methods | B&C | Gauss | JPEG | Moire | Warp | Combined |
|---|---------|-----|-------|------|-------|------|----------|
| 50 | HiDDeN [7] | 69.2% | 67.0% | 60.5% | 68.0% | 69.5% | 56.0% |
| | StegaStamp [11] | 99.2% | 99.9% | 99.9% | 99.9% | **99.9%** | 97.8% |
| | Ours | **99.9%** | **99.9%** | **99.9%** | **99.9%** | 99.7% | **98.8%** |
| 100 | HiDDeN [7] | 64.2% | 64.4% | 56.1% | 61.8% | 63.4% | 52.0% |
| | StegaStamp [11] | 99.0% | 99.8% | 99.8% | 99.7% | 99.0% | 95.7% |

(*continued*)

**Table 3.** (*continued*)

| L | Methods | B&C | Gauss | JPEG | Moire | Warp | Combined |
|---|---------|-----|-------|------|-------|------|----------|
|   | Ours | **99.8%** | **99.9%** | **99.9%** | **99.9%** | **99.4%** | **97.5%** |
| 150 | HiDDeN [7] | 54.9% | 54.5% | 53.4% | 54.2% | 53.6% | 51.0% |
|   | StegaStamp [11] | 98.2% | 99.4% | 99.4% | 99.3% | 99.2% | 95.8% |
|   | Ours | **99.8%** | **99.9%** | **99.9%** | **99.9%** | **99.4%** | **97.6%** |
| 200 | HiDDeN [7] | 50.1% | 48.6% | 50.3% | 49.2% | 49.8% | 51.6% |
|   | StegaStamp [11] | 92.5% | 94.2% | 94.2% | 93.9% | 94.0% | 90.1% |
|   | Ours | **97.5%** | **97.8%** | **97.9%** | **97.8%** | **97.1%** | **94.6%** |

StegaStamp have better robustness than HiDDeN under various attacks. Specifically, if Brightness and Contrast offset is selected to attack the watermarked image, our method is always better than HiDDeN and StegaStamp with the increase of attack intensity, and can achieve the highest bit recovery rate.

Overall, StegaStamp has strong robustness but slightly poor visual quality, whereas HiDDeN has good visual quality but poor robustness. Compared with these two methods, our method can obtain better visual quality and higher robustness simultaneously.

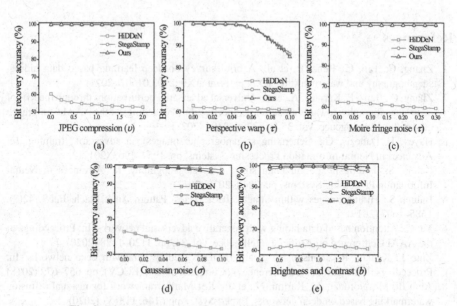

**Fig. 4.** Bit recovery accuracy under different noise intensity attacks.

## 4  Conclusion

In this paper, an adaptive robust watermarking algorithm based on DNNs is proposed (the code is available at https://github.com/Liv2016/Robust-watermarking). The main contributions are as follows.

1) The noise pattern is adaptively generated according to the statistical properties of the carrier image, and thus, after adding it to the carrier image, good visual quality can be easily obtained.
2) With considering a series of attacks such as Perspective warp, adding noise, and JPEG compression, the watermark decoder in our scheme can efficiently resist some potential attacks in the physical world.
3) Experimental results demonstrate that compared with the state-of-the-art algorithms based on DNNs, our method can get better visual quality and higher robustness at the same time. The problem of mutual restriction between visual quality and robustness is solved to some extent.

**Acknowledgments.** This work was supported by the National Natural Science Foun-dation of China (62072481, 61772572), and the Science and Technology Program of Guangzhou, China (202201011587).

## References

1. Zhang, C., Lin, C., Benz, P., et al.: A brief survey on deep learning based data hiding, steganography and watermarking. arXiv preprint arXiv:2103.01607 (2021)
2. Zhang, C., Benz, P., Karjauv, A., et al.: Universal adversarial perturbations through the lens of deep steganography: towards a Fourier perspective. In: Proceedings of the AAAI Conference on Artificial Intelligence, vol. 35, pp. 3296–3304 (2021)
3. Hayes, J., Danezis, G.: Generating steganographic images via adversarial training. In: Advances in Neural Information Processing Systems, pp. 1954–1963 (2017)
4. Baluja, S.: Hiding images in plain sight: deep steganography. In: Advances in Neural Information Processing Systems, pp. 2069–2079 (2017)
5. Baluja, S.: Hiding images within images. IEEE Trans. Pattern Anal. Mach. Intell. **42**(7), 1685–1697 (2019)
6. Yu, C.: Attention based data hiding with generative adversarial networks. In: Proceedings of the AAAI Conference on Artificial Intelligence, vol. 34, pp. 1120–1128 (2020)
7. Zhu, J., Kaplan, R., Johnson, J., et al.: HiDDeN: hiding data with deep networks. In: Proceedings of the European Conference on Computer Vision (ECCV), pp. 657–672 (2018)
8. Ahmadi, M., Norouzi, A., Karimi, N., et al.: ReDMark: framework for residual diffusion watermarking based on deep networks. Expert Syst. Appl. **146**, 113157 (2020)
9. Liu, Y., Guo, M., Zhang, J., et al.: A novel two-stage separable deep learning framework for practical blind watermarking. In: Proceedings of the 27th ACM International Conference on Multimedia, pp. 1509–1517 (2019)
10. Zhang, C., Karjauv, A., Benz, P., et al.: Towards robust data hiding against (JPEG) compression: a pseudo-differentiable deep learning approach. arXiv preprint arXiv:2101.00973 (2020)

11. Tancik, M., Mildenhall, B., Ng, R.: Stegastamp: invisible hyperlinks in physical photographs. In: Proceedings of the IEEE Conference on Computer Vision and Pattern Recognition, pp. 2117–2126 (2020)

12. Wengrowski, E., Dana, K.: Light field messaging with deep photographic steganography. In: Proceedings of the IEEE Conference on Computer Vision and Pattern Recognition, pp. 1515–1524 (2019)

13. Luo, X., Zhan, R., Chang, H., et al.: Distortion agnostic deep watermarking. In: Proceedings of the IEEE Conference on Computer Vision and Pattern Recognition, pp. 13548–13557 (2020)

14. Cui, H., Bian, H., Zhang, W., et al.: Unseencode: invisible on-screen barcode with image-based extraction. In: IEEE Conference on Computer Communications, IEEE INFOCOM 2019, pp. 1315–1323. IEEE (2019)

15. Zhang, C., Benz, P., Karjauv, A., et al.: UDH: universal deep hiding for steganography, watermarking, and light field messaging. In: Advances in Neural Information Processing Systems, pp. 10223–10234 (2020)

16. Abdelnabi, S., Fritz, M.: Adversarial watermarking transformer: towards tracing text provenance with data hiding. In: 2021 IEEE Symposium on Security and Privacy (SP), pp. 121–140. IEEE (2021)

17. Zhou, Z., Siddiquee, M.M.R., Tajbakhsh, N., et al.: UNet++: redesigning skip connections to exploit multiscale features in image segmentation. IEEE Trans. Med. Imaging $39$(6), 1856–1867 (2019)

18. Shen, B., Sethi, I.K., Bhaskaran, V.: DCT domain alpha blending. In: Proceedings 1998 International Conference on Image Processing. ICIP98 (Cat. No. 98CB36269), pp. 857–861. IEEE (1998)

19. Zhang, R., Isola, P., Efros, A., et al.: The unreasonable effectiveness of deep features as a perceptual metric. In: Proceedings of the IEEE Conference on Computer Vision and Pattern Recognition, pp. 586–595 (2018)

20. Huiskes, M.J., Lew, M.S.: The MIR Flickr retrieval evaluation. In: Proceedings of the 1st ACM International Conference on Multimedia Information Retrieval, pp. 39–43 (2008)

21. Russakovsky, O., Deng, J., Su, H., et al.: ImageNet large scale visual recognition challenge. Int. J. Comput. Vis. $115$(3), 211–252 (2015)

22. Kingma, D.P., Ba, J.: Adam: a method for stochastic optimization. arXiv preprint arXiv:1412.6980 (2014)

23. Huynh-Thu, Q., Ghanbari, M.: Scope of validity of PSNR in image/video quality assessment. Electron. Lett. $44$(13), 800–801 (2008)

24. Wang, Z., Bovik, A.C., Sheikh, H.R., et al.: Image quality assessment: from error visibility to structural similarity. IEEE Trans. Image Process. $13$(4), 600–612 (2004)

# Adaptive Despread Spectrum-Based Image Watermarking for Fast Product Tracking

Fei Zhang, Hongxia Wang$^{(\boxtimes)}$, Mingze He, and Jinhe Li

School of Cyber Science and Engineering, Sichuan University, Chengdu 610065, China
{zhangfei,hemingze,jhon_ranble}@stu.scu.edu.cn, hxwang@scu.edu.cn

**Abstract.** With the digitalization of the physical markets, the number of users engaging in e-commerce and shopping online is rapidly increasing. An important application area for digital watermarking is the product tracking scenario. For product tracking scenario, watermarking can be used to provide both product links and copyright protection, so the robustness and extraction efficiency of watermarking are the most important metrics. The auto-convolution function (ACNF) based watermarking scheme is the latest image watermarking that achieves the most comprehensive robustness. However, ACNF watermarking is not resilient in the case of Gauss noise and average filtering. Besides, ACNF watermarking focuses only on robustness and ignores extraction efficiency, and the low efficiency of watermark extraction leads to unpleasant user experience. In this paper, we propose an adaptive despread spectrum-based image watermarking for fast product tracking. For watermark embedding, we construct a low-frequency watermark signal in the spatial domain to enhance the robustness to signal processing attacks. In watermark extraction, our scheme uses discrete wavelet transform (DWT) for image dimensionality reduction and adaptively watermark despread spectrum according to the wavelet decomposition level, which can achieve accurate and fast extraction of watermark. The experimental results demonstrate that our proposed watermarking scheme has superior robustness and extraction efficiency than the existing methods under the same imperceptibility.

**Keywords:** Image watermarking · Robustness · E-commerce

## 1 Introduction

Online shopping has become a daily behavior for the public. Adding invisible website links to product images is an important application direction for digital watermarking [1]. Figure 1 gives an illustration of the digital watermarking technology used in a product tracking scenario. As shown in Fig. 1(a), users usually save (download or screenshot) the picture of the product they want to buy through their mobile devices. As shown in Fig. 1(b), in order to make it easier for users to get the product link directly through the image for accurate purchase, digital watermarking technology can be used to hide the product link

© The Author(s), under exclusive license to Springer Nature Switzerland AG 2023
X. Zhao et al. (Eds.): IWDW 2022, LNCS 13825, pp. 174–189, 2023.
https://doi.org/10.1007/978-3-031-25115-3_12

 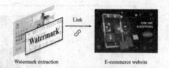

(a) Users obtain the product picture through the online platform

(b) Extract invisible watermark from product image to
link to e-commerce website

**Fig. 1.** Illustration of digital watermarking applied to product tracking scenario.

information into the image in an invisible form. Mobile device can be linked to corresponding e-commerce website through digital watermark.

When watermarking is applied to copyright protection and traceability scenario, it allows for longer execution times since watermark extraction is an offline procedure [2]. Yet, if the watermark is applied to the product tracking scenario, the watermark extraction time is too slow, leading to bad user experience. With the rapid development of mobile terminal camera technology, most of the images taken by mobile phones are currently close to 4K resolution, and most of the mobile phone screens (screenshot images) are close to 1080p resolution. However, most of the watermarking schemes [3–5], which are studied on a standard set of images (512 × 512 resolution), do not take into account the practicality and time efficiency of the algorithms at real image resolutions.

Digital watermarking technology has been extensively studied and many research outcomes have been achieved in the past decades. Divided by embedding domain, watermarking schemes can be classified as spatial domain watermarking and transform domain watermarking. Earlier spatial watermarking includes Least Significant Bit (LSB) [6] and Patchwork [7], but they have very limited robustness. In order to improve the robustness of watermarking, transform domain watermarking has been gradually developed, Discrete Cosine Transform (DCT) [8], Discrete Wavelet Transform (DWT) [9], Duat-Tree Complex Wavelet Transform (DT CWT) [10], etc. are commonly used transform domains. In order to resist geometric attacks, some scholars embed watermarks into geometric invariant moments, Zernike Moments (ZMs) [11], Radial Harmonic Fourier Moments (RHFMs) [12], etc. are commonly used geometric invariant moments. The existing state-of-the-art various types of watermarking have their inherent disadvantages, such as moment-based image watermarking [12] cannot resist cropping attacks and have less efficient algorithms. Transform domain-based watermarking [8] lacks resilience to arbitrary angle rotation attacks. The latest auto-convolution function based watermarking scheme [3] obtains the most comprehensive robustness, which is robust to most common geometric attacks, but it is not robust to average filtering and Gauss noise due to the medium and high frequency propert of its watermarked signal. Auto-convolution function based watermarking is less time efficient in the case of large resolution images, and besides, it needs to know the number of symmetric peaks (i.e., the original image resolution) as side information to accomplish watermark extraction.

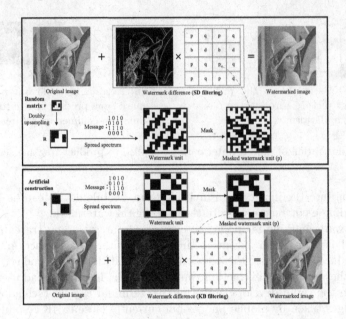

**Fig. 2.** Comparison of the watermark embedding workflow between Ma et al. [3]. (top) and our proposed method (bottom). The key difference is that the proposed method artificially constructs the spread spectrum matrix $R$ at low frequency and introduces a KB filter to generate the watermark difference.

Inspired by this, we propose an adaptive despread spectrum-based image watermarking for fast product tracking. The discrete wavelet transform can be used for image dimensionality reduction to reduce the time for watermark extraction and the low frequency component will have better robustness to signal processing attacks. On the embedding side, we introduce a KB (Ker-Bohme) filter [13] from the adaptive steganography to convolve the original image to determine the pixel-level embedding strength of the watermark, resulting in better imperceptibility. Finally, the watermark is embedded into the image spatial domain in the form of spread spectrum. At the extraction end, we need to perform discrete wavelet transform on the watermarked image in order to achieve fast extraction of the watermarked, and adaptively select the corresponding spread spectrum matrix for watermark despread spectrum according to the transform level. Besides, we also introduce a nonlinear filter at the extraction end to remove the pseudo-peak points generated by the auto-convolution function, so that the algorithm does not need to know the resolution of the original image for watermark extraction. In summary, the contributions of this paper are three-fold:

- We propose to use DWT for image dimensionality reduction and adaptive watermark despread spectrum according to the wavelet decomposition level, which can achieve accurate and fast extraction of watermark.
- We design a nonlinear filter to remove the symmetric peak noise, which allows the watermarking scheme to be applied to various types of images with

different resolutions and to achieve blind extraction without any additional side information about the original image.
- We introduce the KB filter to guide the watermark signal not to embed into clean edges, which improves the imperceptibility of the watermark.

## 2   Motivation

In this section, we first briefly review the work [3], a recent representative auto-convolution function (ACNF) based watermarking scheme, and then point out its inherent limitation. This motivates us to design a watermark fast extraction framework. The embedding process of the scheme in [3] is given in Fig. 2. By applying a key, we generate a bipolar random matrix $r$. Then $r$ is doubly upsampled to obtain spread spectrum matrix $R$. Watermark message is encoded using $R$ to obtain a watermark unit. Watermark unit needs to be masked to form the final masked watermark unit, denoted as p. Flip p until it is the same size as the original image, and then multiply it element by element with the standard deviation filtered image to get the watermark difference. Finally, the watermark difference is directly superimposed into the component Y of original image (YCbCr space) to complete the embedding of the watermark.

As can be seen in Fig. 2, after standard deviation (SD) filtering of the Lena image, the watermark difference outputs high corresponding values at the clean edges. Indeed, both smooth areas and clean edges belong to the sensitive areas of the human eye [14]. Therefore, using standard deviation filtering to determine the watermark strength can cause distortion at the clean edges where the human eye is more sensitive, thus compromising imperceptibility.

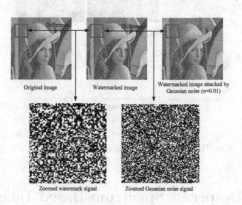

**Fig. 3.** Frequency distribution of Gauss noise signal and watermarked signal of auto-convolution function based watermarking scheme [3].

In watermark extraction, Wiener filtering of the component Y is required to estimate the watermark signal. Then, the auto-convolution of the watermark signal is calculated to obtain the position of each watermark unit. Finally, each watermark unit is demasked and despread spectrum to complete the watermark extraction. Due to the complexity of the extraction process, when the scheme is applied to large resolution images, the time complexity of the auto-convolution algorithm increases and the watermark extraction process becomes very slow. In addition, [3] needs to know the resolution of the original image in order to complete the filtering of symmetric peaks during the binarization of the auto-convolution signal. However, the resolution of products images in real-life applications is varied, and the condition that the watermark extraction requires the resolution of the original image limits the practical application of the watermark. More details related to the watermark extraction can be found in Sect. 4 of the work [3].

In our experiments, we find that the scheme [3] has poor robustness against Gauss noise and average filtering. The frequency distribution of the watermark signal generated by [3] as well as the Gauss noise signal is given in Fig. 3. It can be seen that the frequency of the watermark signal belongs to the medium-high frequency. Gauss noise is closer to the frequency distribution of the watermark signal, which can cause serious interference to the watermark signal. Average filtering is often used to remove Gauss noise. Therefore, [3] is poorly resilient to Gauss noise and average filtering attacks. To this end, we propose an adaptive despread spectrum-based image watermarking in the next section to address the inherent limitations of the scheme in [3].

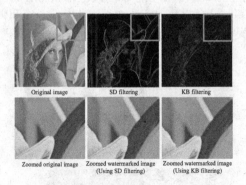

**Fig. 4.** Comparison of the imperceptibility of the scheme [3] using SD (Standard deviation) filter and KB (Ker-Bohme) filter, respectively.

# 3   Adaptive Despread Spectrum-Based Image Watermarking

## 3.1   Watermark Embedding

Figure 2 shows the workflow of the proposed watermark embedding process. As described in Sect. 2, clean edges are not suitable for embedding high-intensity watermark signal. Assume that component Y of the original image $M$ is $I$ of size $N \times M, I_{i,j} \in [0, 255], i = 1, 2, ..., N, j = 1, 2, ..., M$. Here, we use the KB (Ker-Bohme) filter [13] commonly used in adaptive steganography to compare the SD (standard deviation) filter in [3]. KB filter as a form as:

$$KB = \begin{bmatrix} -1 & 2 & -1 \\ 2 & -4 & 2 \\ -1 & 2 & -1 \end{bmatrix}$$

Figure 4 gives the watermark intensity matrices obtained using the KB filter and the SD filter, respectively, and a comparison of the imperceptibility of the watermark embedded using the scheme in [3]. Here, we control other factors constant and only change the filter. It can be seen that compared to the watermarked image obtained by the SD filter, the watermarked image obtained by the KB filter does not cause distortion at the clean edges perceived by the human eye, and the imperceptibility of the watermark is better. Thus, the watermark intensity matrix $s$ is determined in our scheme using the KB filter:

$$s = log_2(|I \otimes KB|). \tag{1}$$

where the notation $\otimes$ stands for the mirror-padded convolution. For the elements in $s$, we restrict the minimum value to 1.

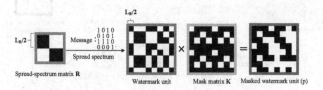

**Fig. 5.** Artificially constructing the spread spectrum matrix as well as the mask matrix makes the watermark signal more low frequency.

As shown in Fig. 5, we manually constructed the spread spectrum matrix $R$ of size $L_R$, with white denoting pixels value $+1$ and black denoting pixels value $-1$. After spreading the message matrix to form a watermark unit, it is masked by a mask matrix $K$ of size $L_K$. The potential motivation behind this strategy is that we can artificially control the frequency distribution of the watermarked signal in the spatial domain. Here, we suggest that $L_R \geq 8$ to make the signal frequency distribution more low-frequency and thus enhance its robustness against Gauss noise and average filtering attacks.

After getting the masked watermark unit p, we flip it until it is the same size as the original image. The final watermark difference is obtained using the watermark intensity $s$. The watermark difference is superimposed onto the original image to complete the embedding of the watermark.

## 3.2   Watermark Extraction

In general, the embedding and extraction domain of the watermark should be consistent. However, in our scheme, although the watermark is embedded into the spatial domain, we choose to extract the watermark signal in the wavelet domain. The potential motivations are as follows: first, the discrete wavelet transform (DWT) can reduce the dimensionality of the image, which will improve the efficiency of watermark extraction. Second, our watermark signal is embedded in the original image in the form of low frequency, it will map to the low frequency sub-band (LL) of DWT, and theoretically we can adjust the spread spectrum matrix and mask matrix in different wavelet decomposition levels to achieve the extraction of the watermark signal in the spatial domain.

After lossy channel transmission, we obtain the degraded watermarked image $M^*$. The component Y of $M^*$ is denoted as $I^*$ of size $N^* \times M^*$. Similarly, $I^*_{i,j} \in [0, 255], i = 1, 2, ..., N^*, j = 1, 2, ..., M^*$. Note that $N^*$, $M^*$ are not necessarily equal to $N, M$, respectively, as the image suffer from unknown distortion. Figure 6 shows the proposed watermark extraction process. We follow the steps below for watermark extraction.

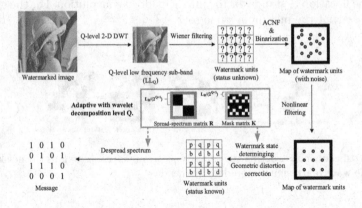

**Fig. 6.** The workflow of the proposed watermark extraction.

**Step 1 Discrete Wavelet Transform:** We need to perform DWT on $I^*$. This work selects Haar wavelet to do DWT for images, because it can make a watermark algorithm more robust to attacks than others [15]. An image can be divided into a lower resolution approximation image ($LL_1$) and three detailed components, vertical ($LH_1$), diagonal ($HH_1$) and horizontal ($HL_1$). The dimensions of $LL_1, LH_1, HL_1$ and $HH_1$ are $(N^*/2) \times (M^*/2)$.

Generally speaking, the $LL_1$ component can be DWT again to obtain the high-level decomposition results, i.e., $LL_2, LH_2, HH_2$ and $HL_2$. As the watermark signal in the spatial domain is low frequency, we only consider watermark extraction in the wavelet low frequency sub-band $LL_Q, Q = 1, 2, 3, \cdots$. The larger the $Q$ value, the higher the watermark extraction efficiency. Besides, we need to adjust the spread spectrum matrix $R$ and the mask matrix $K$ adaptively according to the wavelet decomposition level $Q$. Specifically, the extraction on $LL_Q$ requires performing a scaling of $1/2^Q$ times for $R$ and $K$. The scaling process uses nearest-neighbor interpolation.

**Step 2 Wiener Filtering:** After determining the DWT decomposition level $Q$, we obtained the $LL_Q$, and the corresponding $R$ and $K$. Here we use the common Wiener filter for the estimation of the watermark signal $\hat{W}$, i.e.

$$\hat{W} = (LL_Q - \mu_{LL_Q})\frac{max(\sigma^2_{LL_Q}, mean(\sigma^2_{LL_Q}))}{\sigma^2_{LL_Q}}. \tag{2}$$

where $\mu_{LL_Q}$ and $\sigma^2_{LL_Q}$ are the local mean and local variance of $LL_Q$, respectively. $mean(\cdot)$ indicates the mean value of the matrix. $max(\cdot)$ denotes taking the maximum value.

**Step 3 Compute the Auto-convolution to Obtain the Watermark Units Map:** To determine the location of the watermarked units, we take the auto-convolution function (ACNF) to calculate the symmetry $S$:

$$S = \mathcal{D}(IFFT(FFT(\hat{W}_p))FFT(\hat{W}_p)). \tag{3}$$

where $FFT$ and $IFFT$ stand for the fast Fourier transform and the inverse fast Fourier transform, respectively. $\mathcal{D}$ is a downsampling function that scales its input matrix to the half size. $\hat{W}_p$ is $\hat{W}$ zero-padding to double the original size. To separate symmetrical peaks, [3] use an adaptive threshold to obtain a watermark unit map $A$:

$$A = \begin{cases} 1, if \ S > \mu_S + \beta\sigma_S \\ 0, \ otherwise \end{cases} \tag{4}$$

where $\mu_S$ and $\sigma_S$ denote the local average and the standard deviation of $S$. $\beta$ is an empirical coefficient that is set from 3.0 to 4.3. However, the actual calculation of $A$ in [3] requires a rough idea of the number of original symmetric peaks (i.e., the resolution of the original image) for the best extraction, as detailed in the author's publicly available code [16]. However, in real life, the resolution of images is various. We wish to extract the watermark with as less side information as possible in order to improve the practicality of the scheme. In this paper, a nonlinear filter is proposed that can remove pseudo-peak points (noise) to obtain a clean map of watermark units without the need to know the number of original peak points.

**Step 4 Nonlinear Filtering:** We use the method in [3] as a basis to search for as many peak points as possible, i.e., we assume that the original number of peak points is the maximum within a reasonable range. Many pseudo-peak points (noise) appear after the execution of Eq. (4). We assume that the true peak point is maximum in the local area. Therefore, we design a nonlinear filter as follows to obtain the de-noised watermark units map $A'$:

$$A' = \begin{cases} 1, if \ max(Z) = S \ \& \ A = 1 \\ 0, \ otherwise \end{cases} \tag{5}$$

where $Z$ denotes the local neighborhood of size $(L_K/2^{Q+2}-1) \times (L_K/2^{Q+2}-1)$. Here $L_K$ is the size of the original mask matrix $K$, i.e. the size of the watermark unit. The size of the local region $Z$ is chosen to be $1/2$ the size of the watermark unit in the current DWT decomposition level, i.e., $L_K/2^{Q+2}$. The final area range is performed -1 so that the true peak point is not deleted when the 0.5 times scaling attack occurs.

An example of the effect of nonlinear filtering for de-noising is given in Fig. 7. We need to estimate as many peak points as possible and then use a nonlinear filter to get clean peak points. This process does not require knowledge of the resolution of the original image, enhancing the practicality of existing scheme. Now, we can estimate the geometric distortion parameters of the image based on $A'$ and thus correct the geometric distortion.

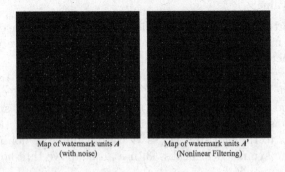

Map of watermark units $A$
(with noise)

Map of watermark units $A'$
(Nonlinear Filtering)

**Fig. 7.** Demonstration of de-noising effect of nonlinear filter.

**Step 5 Watermark State Determining:** We first need to obtain the spread spectrum matrix $R$ and the mask matrix $K$ adjusted according to the wavelet decomposition level Q. After recovering the geometric distortion, we can obtain a partial set of watermark units, expressed as $W_{units} = \{W_i, i \in \mathbb{N}^+\}$, and the number of watermark units is $|W_{units}|$. As shown is Fig. 8, there are 8 possible states of the watermark unit $W_i$. We use $K$ demask it to get $W_i^r$ and construct a null hypothesis as follows:

$$H_0 : W_i \text{ is } \textbf{not} \text{ state 1.}$$

If $H_0$ is false and the correlation between $W_i^r$ and the spread spectrum matrix $R$ is higher, we divide $W_i^r$ into non-overlapping blocks according to the size of $R$, expressed as $(W_i^r)_{n,m}$, where $n, m \in [1, \sqrt{L}]$. $L$ is the length of the watermark. Then the correlation $Y$ between $W_i^r$ and $R$ can be calculated as:

$$Y = \sum_n \sum_m |C((W_i^r)_{n,m}, R)| \tag{6}$$

where $C(\cdot)$ represent the inner product. Assume that the set of possible null hypothesis is $\Lambda$:

$$\Lambda = \{H_0 : \boldsymbol{W}_i \text{ is } \mathbf{not} \text{ state } z, z \in [1,8]\}$$

Therefore, we take the null hypothesis with the highest correlation $Y$ in the set $\Lambda$ to reject it and determine the final watermark state.

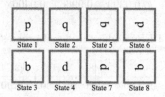

**Fig. 8.** Eight possible watermark unit states.

**Step 6 Despread Spectrum:** Once the state of the single watermark unit $\boldsymbol{W}_i$ has been determined, watermark extraction can be performed:

$$\hat{w}_i(n,m) = \begin{cases} 1, & if \ C((\boldsymbol{W}_i^r)_{n,m}, \boldsymbol{R}) \geq 1 \\ 0, & otherwise \end{cases} \tag{7}$$

To improve robustness, we traverse $\boldsymbol{W}_{units} = \{\boldsymbol{W}_i, i \in \mathbb{N}^+\}$ and give the final extracted message $\hat{w}$:

$$\hat{w}(n,m) = \begin{cases} 1, & if \ \sum_i \hat{w}_i(n,m) \geq |\boldsymbol{W}_{units}|/2 \\ 0, & otherwise \end{cases} \tag{8}$$

## 4 Experimental Evaluation

### 4.1 Experimental Setup

In our experiments, we choose the wavelet decomposition level $Q = 1$ and the spread spectrum matrix $\boldsymbol{R}$ size $L_R = 8$. For all watermarking schemes in the comparison experiments, their message length $L$ is uniformly set to 64 bits. As described in Sect. 1, the current state-of-the-art watermarking schemes are DCT domain spread spectrum-based watermarking [8], moment-based watermarking [12], and auto-convolution function based watermarking [3]. We use SSAES (**S**pread **s**pectrum scheme with **a**daptive **e**mbedding **s**trength) [8], RHFMs-FFT (**R**adial **h**armonic **F**ourier **m**oments based on **f**ast **F**ourier transform) [12], and ACNF (**A**uto-**c**onvolution **f**unction) [3] to represent the above three watermarking schemes in order to concisely illustrate the experimental results.

The test host images consisted of 120 randomly selected color images from the USC-SIPI image database [17] and Yahoo's image sharing website [18], 60 of size $512 \times 512$ and another 60 of size $1920 \times 1080$. Besides, to verify that the watermarking scheme can be applied to various image formats, we also adjust the format of the image dataset. We adjusted the format of sixty $512 \times 512$ images so that they included 20 **jpg** images, 20 **bmp** images and 20 **png** images. Similar

**Table 1.** Average PSNR and SSIM of watermarked images generated by different schemes.

| Scheme | [8] | [12] | [3] | Proposed |
|---|---|---|---|---|
| PSNR (dB) | 41.4965 | 41.5730 | 41.8620 | 41.8765 |
| SSIM | 0.9399 | 0.9966 | 0.9406 | 0.9717 |

**Table 2.** BERs under the signal process attacks.

| Attack types | [8] | [12] | [3] | Proposed |
|---|---|---|---|---|
| JPEG compression 80% | 0.003 | 0.001 | **0.000** | **0.000** |
| JPEG compression 60% | 0.003 | 0.001 | **0.000** | 0.002 |
| JPEG compression 40% | 0.006 | **0.002** | 0.018 | 0.012 |
| Gauss noise 0.0001 | 0.002 | 0.001 | **0.000** | **0.000** |
| Gauss noise 0.001 | 0.002 | 0.002 | **0.000** | **0.000** |
| Gauss noise 0.01 | **0.007** | 0.088 | 0.118 | 0.046 |
| Salt & pepper noise 0.001 | 0.002 | 0.001 | **0.000** | **0.000** |
| Salt & pepper noise 0.01 | 0.003 | 0.003 | **0.000** | 0.001 |
| Salt & pepper noise 0.02 | 0.005 | 0.006 | **0.000** | 0.003 |
| Average filtering $3 \times 3$ | 0.003 | 0.001 | **0.000** | 0.001 |
| Average filtering $5 \times 5$ | 0.008 | **0.001** | 0.035 | 0.003 |
| Average filtering $7 \times 7$ | 0.495 | **0.001** | 0.410 | 0.004 |
| Laplacian filtering $3 \times 3$ | 0.002 | 0.071 | **0.000** | **0.000** |
| Laplacian filtering $5 \times 5$ | 0.002 | 0.071 | **0.000** | **0.000** |
| Laplacian filtering $7 \times 7$ | 0.002 | 0.071 | **0.000** | **0.000** |
| Contrast change −30% | 0.002 | 0.477 | 0.001 | **0.000** |
| Contrast change −50% | 0.002 | 0.605 | **0.000** | **0.000** |
| Contrast change +30% | 0.004 | 0.408 | **0.001** | 0.005 |
| Contrast change +50% | **0.006** | 0.480 | 0.015 | 0.045 |
| Saturation change −30% | 0.002 | 0.025 | **0.000** | 0.004 |
| Saturation change −50% | 0.002 | 0.055 | **0.000** | **0.000** |
| Saturation change +30% | 0.002 | 0.022 | **0.000** | **0.000** |
| Saturation change +50% | 0.002 | 0.062 | **0.000** | **0.000** |
| Luminance change −30% | 0.002 | 0.001 | **0.000** | **0.000** |
| Luminance change −50% | 0.002 | 0.001 | **0.000** | **0.000** |
| Luminance change +30% | 0.002 | 0.001 | **0.000** | **0.000** |
| Luminance change +50% | 0.002 | 0.001 | **0.000** | **0.000** |
| Chroma change −30% | 0.002 | 0.112 | **0.000** | 0.001 |
| Chroma change −50% | 0.002 | 0.160 | **0.000** | **0.000** |
| Chroma change +30% | 0.003 | 0.117 | **0.000** | 0.001 |
| Chroma change +50% | 0.002 | 0.093 | **0.000** | **0.000** |

**Fig. 9.** Top row: Host images; Bottom row: Watermarked images generated by the proposed scheme.

formatting has been done for 1080p images. A partial example of the host images generated by this scheme and its corresponding watermarked images are given in Fig. 9. The average peak signal-to-noise ratio (PSNR) of 120 watermarked images is 41.8765 dB. The robustness of the scheme is evaluated by the bit error rate (BER) of the watermarked image at the corresponding distortion. In the comparison experiments, the other watermarking schemes take the same experimental setup. The average PSNR and structural similarity index (SSIM) of the four watermarking schemes are given in Table 1. The average PSNR values of the watermarked images generated by the four different watermarking schemes were all set to the same level of $41.6 \pm 0.3$ dB.

## 4.2   Robustness to Signal Processing Attacks

In this section, we compare the resilience of SSAES [8], FFT-RHFMs [12], and ACNF [3] to signal processing attacks. Attack types include JPEG compression, Gauss noise, Salt & Pepper noise, Average filtering (image blurring), Laplacian filtering (image sharpening) and Contrast/Saturation/Luminance/Chroma change.

The average BER of the four schemes against different signal processing attacks is listed in Table 2. In general, BER $\leq 0.050$ is an acceptable robustness performance. It can be seen that the overall robustness of SSAES [8] watermarking is excellent, but poor when suffering from average filtering ($7 \times 7$) attack. RHFMs-FFT [12] watermark is poorly resilient to contrast change. ACNF watermarking [3] cannot resist Gauss Noise (0.01) attack and average filtering ($7 \times 7$) attack, while our scheme solves the shortcomings of ACNF scheme well. Due to the artificial control that we keep the spatial watermark signal at low frequency, which enhances its resilience to Gauss noise and average filtering attacks. It can be seen that our scheme has better robustness against various types of signal processing attacks and can resist common types of signal processing attacks.

## 4.3   Robustness to Geometric Attacks

In this section, we compare the resilience of SSAES [8], FFT-RHFMs [12], and ACNF [3] to geometric attacks. Attack types include Cropping, Insert, Adding boundary, Scaling, Flipping, Rotation and Aspect ration change.

The average BER of the four schemes against different geometric attacks is listed in Table 3. SSAES [8] watermarking is not resistant to geometric attacks that cause resolution changes, such as cropping and adding boundary. Besides, it is not resilient to rotation attacks at arbitrary angles (not multiples of 90). Whereas, FFT-RHFMs [12] watermark is not resistant to cropping, adding boundary and aspect ration change attacks due to the lack of synchronization capability. In contrast, the robustness of ACNF [3] watermarking is more comprehensive. The robustness of our proposed scheme is closer to that of the ACNF scheme and can resist common geometric attacks.

Table 3. BERs under the geometric attacks.

| Attack types | [8] | [12] | [3] | Proposed |
|---|---|---|---|---|
| Cropping 10% | N/A | N/A | **0.000** | **0.000** |
| Cropping 30% | N/A | N/A | **0.000** | 0.001 |
| Cropping 50% | N/A | N/A | **0.000** | 0.001 |
| Insert 10% | 0.003 | 0.281 | **0.000** | 0.001 |
| Insert 30% | 0.003 | 0.490 | **0.000** | 0.002 |
| Insert 50% | 0.014 | 0.489 | **0.001** | 0.050 |
| Adding boundary 10% | N/A | N/A | **0.000** | 0.001 |
| Adding boundary 30% | N/A | N/A | **0.000** | **0.000** |
| Adding boundary 50% | N/A | N/A | 0.192 | **0.043** |
| Scaling 0.5 | 0.002 | 0.001 | **0.000** | 0.001 |
| Scaling 0.7 | 0.002 | 0.001 | **0.000** | 0.009 |
| Scaling 0.9 | 0.002 | 0.001 | **0.000** | 0.010 |
| Scaling 1.1 | 0.002 | 0.001 | **0.000** | 0.006 |
| Scaling 1.3 | 0.002 | **0.001** | 0.008 | 0.006 |
| Scaling 2.0 | 0.002 | **0.001** | 0.067 | **0.001** |
| Horizontal flipping | 0.002 | 0.001 | **0.000** | **0.000** |
| Vertical flipping | 0.002 | 0.001 | **0.000** | 0.001 |
| Rotation −10° | N/A | 0.001 | **0.000** | **0.000** |
| Rotation −20° | N/A | 0.001 | **0.000** | 0.042 |
| Rotation −30° | N/A | 0.001 | **0.000** | 0.024 |
| Rotation −90° | 0.002 | 0.001 | **0.000** | **0.000** |
| Rotation +10° | N/A | 0.001 | **0.000** | **0.000** |
| Rotation +20° | N/A | 0.001 | **0.000** | 0.029 |

(*continued*)

**Table 3.** (*continued*)

| Attack types | [8] | [12] | [3] | Proposed |
|---|---|---|---|---|
| Rotation +30° | N/A | **0.001** | 0.057 | 0.009 |
| Rotation +90° | 0.002 | 0.001 | **0.000** | **0.000** |
| Aspect ration change (0.9 × 1.1) | 0.002 | N/A | **0.000** | 0.007 |
| Aspect ration change (0.8 × 1.2) | **0.002** | N/A | 0.003 | 0.005 |
| Aspect ration change (0.7 × 1.3) | 0.003 | N/A | **0.000** | 0.006 |
| Aspect ration change (1.1 × 0.9) | 0.002 | N/A | **0.000** | 0.011 |
| Aspect ration change (1.2 × 0.8) | 0.002 | N/A | **0.000** | 0.007 |
| Aspect ration change (1.3 × 0.7) | 0.003 | N/A | **0.000** | 0.024 |

**Table 4.** Comparison of the average embedding time (seconds per image) of different watermarking schemes.

| Resolution | 512 × 512 | 1024 × 1024 | 2048 × 2048 | 4096 × 4096 |
|---|---|---|---|---|
| [8] | 0.27 | 0.80 | 3.45 | 13.28 |
| [12] | 7.79 | 34.32 | 140.17 | 693.03 |
| [3] | **0.16** | 0.68 | **2.76** | 11.04 |
| Proposed | 0.17 | **0.66** | 2.79 | **11.01** |

**Table 5.** Comparison of the average extraction time (seconds per image) of different watermarking schemes.

| Resolution | 512 × 512 | 1024 × 1024 | 2048 × 2048 | 4096 × 4096 |
|---|---|---|---|---|
| [8] | **0.09** | **0.16** | **0.46** | **1.43** |
| [12] | 2.03 | 8.62 | 37.35 | 286.15 |
| [3] | 0.70 | 2.00 | 6.91 | 40.08 |
| Proposed ($Q = 1$) | 0.26 | 0.61 | 2.16 | 4.66 |
| Proposed ($Q = 2$) | 0.22 | 0.46 | 1.54 | 2.75 |

## 4.4  Computational Cost

We test the running time on a PC with Intel(R) Core(TM) i7-9700K CPU and 32 GB RAM. We scale 60 512 × 512 images in the image dataset with parameters 1.0, 2.0, 4.0, and 8.0 to generate 60 images of 512 × 512 size, 60 images of 1024 × 1024 size, 60 images of 2048 × 2048 size, and 60 images of 4096 × 4096 size, respectively. We use the obtained 240 images to test the watermark embedding and extraction efficiency of the four schemes.

Table 4 gives a comparison of the average embedding times of the four watermarking schemes on different resolution images. It can be seen that the embedding efficiency of our scheme and ACNF [3] watermark are closer, and have higher embedding efficiency compared to SSAES [8] and FFT-RHFMs [12].

Table 5 gives a comparison of the average extraction times of the four watermarking schemes on different resolution images. It can be seen that the proposed scheme has a significant improvement in extraction efficiency compared to the ACNF [3] watermarking scheme as the Q value increases, and the efficiency is close to that of SSAES. Obviously, our extraction efficiency is greatly improved compared with the original ACNF [3] scheme, and the larger the Q value, the greater the extraction efficiency improvement. It is thanks to DWT that reduces the image dimensionality and reduces the time complexity of the watermark extraction algorithm. If we consider the inherent shortcomings of SSAES [8] watermarking in terms of robustness, our scheme is undoubtedly a better choice for product tracking scenario.

## 5    Conclusion

In this paper, we propose an adaptive despread spectrum-based image watermarking for fast product tracking, which has the core idea of using DWT to reduce the dimensionality of images in order to improve the extraction efficiency of watermarks. The proposed scheme can adaptively adjust the spread spectrum matrix and the mask matrix with the wavelet decomposition level, so that the extraction of low-frequency watermarked signals in the spatial domain can be achieved in the wavelet domain. The design of nonlinear filters improves the practicality of watermarking, allowing the proposed scheme to process images of different resolutions in batch. Besides, we also use the KB filter commonly used in adaptive steganography to guide the watermark without embedding to clean edges, improving the subjective imperceptibility of the watermark. Extensive experimental results demonstrate that our proposed scheme improves the robustness of existing watermarking schemes and substantially improves the extraction efficiency of watermark. In our future work, we will focus on further improving the embedding and extraction efficiency of watermarking schemes on extra-large resolution images.

**Acknowlededgment.** This work was supported in part by the National Natural Science Foundation of China (NSFC) under Grants 62272331 and 61972269, and Sichuan Science and Technology Program under Grant 2022YFG0320.

## References

1. Fang, H., et al.: Deep template-based watermarking. IEEE Trans. Circuits Syst. Video Technol. **31**(4), 1436–1451 (2020)
2. Su, P., Kuo, T., Li, M.: A practical design of digital watermarking for video streaming services. J. Vis. Commun. Image Represent. **42**, 161–172 (2017)
3. Ma, Z., Zhang, W., Fang, H., Dong, X., Geng, L., Yu, N.: Local geometric distortions resilient watermarking scheme based on symmetry. IEEE Trans. Circuits Syst. Video Technol. **31**(12), 4826–4839 (2021)
4. Fang, H., Zhang, W., Zhou, H., Cui, H., Yu, N.: Screen-shooting resilient watermarking. IEEE Trans. Inf. Forensics Secur. **14**(6), 1403–1418 (2018)

5. Kang, X., Huang, J., Zeng, W.: Efficient general print-scanning resilient data hiding based on uniform log-polar mapping. IEEE Trans. Inf. Forensics Secur. **5**(1), 1–12 (2010)

6. Van Schyndel, R., Tirkel, A., Osborne, C.: A digital watermark. In: Proceedings of 1st International Conference on Image Processing, vol. 2, pp. 86–90. IEEE (1994)

7. Bender, W., Gruhl, D., Morimoto, N., Lu, A.: Techniques for data hiding. IBM Syst. J. **35**(3–4), 313–336 (1996)

8. Huang, Y., Niu, B., Guan, H., Zhang, S.: Enhancing image watermarking with adaptive embedding parameter and PSNR guarantee. IEEE Trans. Multimedia **21**(10), 2447–2460 (2019)

9. Liu, J., Tu, Q., Xu, X.: Quantization-based image watermarking by using a normalization scheme in the wavelet domain. Information **9**(8), 194 (2018)

10. Huan, W., Li, S., Qian, Z., Zhang, X.: Exploring stable coefficients on joint subbands for robust video watermarking in DT CWT domain. IEEE Trans. Circuits Syst. Video Technol. **32**(4), 1955–1965 (2021)

11. Xin, Y., Liao, S., Pawlak, M.: Circularly orthogonal moments for geometrically robust image watermarking. Pattern Recogn. **40**(12), 3740–3752 (2007)

12. Chun, W., Xing, W., Zhi, X.: Geometrically invariant image watermarking based on fast radial harmonic Fourier moments. Signal Process. Image Commun. **45**, 10–23 (2016)

13. Li, B., Wang, M., Huang, J., Li, X.: A new cost function for spatial image steganography. In: 2014 IEEE International Conference on Image Processing (ICIP), pp. 4206–4210. IEEE (2014)

14. Holub, V., Fridrich, J., Denemark, T.: Universal distortion function for steganography in an arbitrary domain. EURASIP J. Inf. Secur. **2014**(1), 1–13 (2014). https://doi.org/10.1186/1687-417X-2014-1

15. Liu, Q., Yang, S., Liu, J., Xiong, P., Zhou, M.: A discrete wavelet transform and singular value decomposition-based digital video watermark method. Appl. Math. Model. **85**, 273–293 (2020)

16. Source code of Ref. 3. https://github.com/CirnoGiovanna/LGDR_watermark/

17. The USC-SIPI image database. http://sipi.usc.edu/database/

18. Yahoo's image sharing website. https://www.flickr.com/

# Reversible Data Hiding via Arranging Blocks of Bit-Planes in Encrypted Images

Yingying Sun, Guoxiong Xie, Guijin Fan, Chunqiang Yu, and Zhenjun Tang[✉]

Guangxi Key Lab of Multi-source Information Mining and Security,
Guangxi Normal University, Guilin 541004, China
tangzj230@163.com

**Abstract.** Reversible data hiding in encrypted images (RDHEI) is a useful technique for protecting data security, but most RDHEI methods do not make a satisfied embedding performance yet. To address this, we propose a novel RDHEI method via arranging blocks of bit-planes to vacate more room. In the proposed RDHEI method, the eight bit-planes of prediction error image are divided into non-overlapping blocks, which are classified into two kinds: uniform block (UB) and non-uniform block (NUB). To improve the embedding performance, we not only use the UBs but also exploit the NUBs to vacate room. Specifically, the NUB is further divided into compressible block or non-compressible block according to its detailed bits, where the compressible block is used to vacate room. As more blocks are used for data embedding, the proposed RDHEI method reaches a high embedding performance. Many experiments are done to validate performances of the proposed RDHEI method. The results show that the embedding rates of the proposed RDHEI method on the BOSS-base, BOWS-2, and UCID databases are 3.9773, 3.8800, and 3.1527 bpp, respectively. Comparison illustrates that the proposed RDHEI method outperforms some state-of-the-art RDHEI methods in terms of embedding rate.

**Keywords:** Reversible data hiding · Image encryption · Block arrangement · Embedding rate

## 1 Introduction

With the rapid development of cloud services and 5G networks, Internet communication has become more and more convenient and the amount of data uploaded to the cloud is increasing. Meanwhile, secure data transmission has become an urgent problem to be solved. Reversible data hiding in encrypted images (RDHEI) [1–6] is an effective technique for protecting data security. It embeds secret data into carrier images by making slight modifications. Moreover, it can extract the secret data and exactly recover the carrier images without any error. These advantages make RDHEI suitable for many applications, such as military applications, medical applications and judicial applications. As digital

© The Author(s), under exclusive license to Springer Nature Switzerland AG 2023
X. Zhao et al. (Eds.): IWDW 2022, LNCS 13825, pp. 190–204, 2023.
https://doi.org/10.1007/978-3-031-25115-3_13

images are widely used in many fields, many researchers have designed various RDHEI methods for data security. Nowadays, the existing RDHEI methods can be divided into two types according to the order of reserving room and image encryption: Reserving room before encryption (RRBE) [7–11] and vacating room after encryption (VRAE) [12–16]. Some of their typical methods are introduced in the below paragraphs.

(1) *RRBE based RDHEI methods*: An early work of this kind methods was given by Ma *et al.* [7]. They used a traditional reversible data hiding (RDH) method to reserve room before encryption and achieved data extraction and image recovery without error. In another work, Zhang *et al.* [8] presented a novel RDHEI method. They estimated some pixels before encryption and embedded secret data into the estimated errors. In addition, they designed a special encryption scheme to encrypt those estimated errors. Their data extraction and image decryption are independent. To vacate more room, Wu *et al.* [9] exploited adaptive prediction-error labeling and encryption to design two methods: Exposing the shuffled labels in encrypted images and performing the encryption on the labels. As the binary image do not have much redundancy, Ren *et al.* [10] divided the binary image into uniform blocks and non-uniform blocks, and embedded the secret data into these blocks. To increase the reserved room before image encryption, Yin *et al.* [11] improved the method given by Ren *et al.* [10] by using pixel prediction and bit-plane rearrangement. Unlike the previous methods, they divided bit-planes into uniform blocks and non-uniform blocks, and reordered these blocks to reserve embeddable room.

(2) *VRAE based RDHEI methods*: A typical method was introduced by Zhang [12] in 2011. He used a stream cipher to encrypt image, and modified some encrypted data to embed secret data into the image. To increase the embedding rate, Tang *et al.* [13] designed a block-based encryption method and proposed a differential compression (DC) technique to compress the encrypted image. The DC technique can vacate more room for embedding secret data. In another work, Tang *et al.* [14] exploited a novel technique called adaptive prediction error coding to design a new RDHEI method. As the adaptive prediction error coding can vacate more room for data embedding, this RDHEI method achieves a high embedding rate. Recently, Yu *et al.* [15] exploited adaptive difference recovery to design a new technique of data hiding and used it make an efficient RDHEI method. They preserved spatial redundancy within encrypted blocks and conducted data embedding by the adaptive difference recovery. Gao *et al.* [16] combined the most significant bit prediction and error embedding to construct a novel RDHEI method. They classified blocks into three types: flag blocks, error blocks and message blocks. In this method, the flag blocks are used to distinguish error blocks and message blocks, and the message blocks are used to embed data.

Many RDHEI methods are reported in the past years and some of them make good performances in the security and the embedding rate. However, the embedding performance of most RDHEI methods are not satisfied yet because the demand of hiding secret data is rapidly increasing. In this paper, we propose a novel RDHEI method via arranging blocks of bit-planes to vacate more room. In the proposed RDHEI method, the eight bit-planes of prediction error (PE) image are divided into non-overlapping blocks, which are classified into two kinds: uniform block (UB) and non-uniform block (NUB). To improve the embedding performance, we not only use the UBs but also exploit the NUBs to vacate room. Specifically, the NUB is further divided into compressible block (CB) or non-compressible blocks (NCB) according to its detailed bits, where the CB is used to vacate room. As more blocks are used for data embedding, the proposed RDHEI method reaches a high embedding performance. Many experiments on three open databases are done to validate performances of the proposed RDHEI method. The results show that the embedding rates of the proposed RDHEI method on the BOSSbase, BOWS-2, and UCID databases are 3.9773, 3.8800, and 3.1527 bpp, which are bigger than those of the state-of-the-art methods [9,11,15,16].

The rest of this paper is organized as follows. Section 2 explains the proposed RDHEI method in detail. Section 3 discusses the experimental results, and Sect. 4 concludes this paper.

## 2   Proposed Method

The framework of the proposed RDHEI method contains three stages, as shown in Fig. 1. In the first stage, the content owner calculates PE image, decomposes the PE image into eight bit-planes, divides each bit-plane into non-overlapping blocks (i.e., UBs and NUBs), and arranges these blocks to vacate room. Then the content owner encrypts the pre-processed image with an encryption key. The output of the first stage is the encrypted image with the reserved room. In the second stage, the data hider embeds secret data into the reserved room of the encrypted image by a data hiding key. After that, the marked encrypted image is obtained. In the third stage, the receiver can extract secret data or recover image with the guide of auxiliary information according to the available knowledge of the data hiding key and encryption key. Details of the proposed RDHEI method are explained as follows.

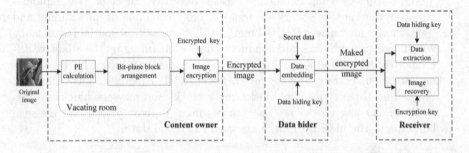

**Fig. 1.** Framework of the proposed RDHEI method

## 2.1   Vacating Room

The procedure of vacating room includes two parts: PE calculation, and bit-plane block arrangement. Their details are explained as follows.

**(1) PE Calculation.** Suppose that the size of input image is $M \times N$ and its pixel in the $i$-th row and $j$-th column is denoted as $x(i,j) \in [0, 255]$, where $1 \le i \le M$ and $1 \le j \le N$. During the predicted image calculation, the pixels located in the first row/column remain unchanged as reference pixels, and the predicted values of other pixels are obtained by their two adjacent pixels. Here, the predicted value $p(i,j)$ of the pixel $x(i,j)$ can be calculated by the Eq. (1).

$$p(i,j) = \begin{cases} x(i,j), & \text{If } i = 1 \parallel j = 1 \\ \lfloor 0.5x(i,j-1) + 0.5x(i-1,j) \rfloor, & \text{Otherwise} \end{cases} \tag{1}$$

where $\lfloor \cdot \rfloor$ represents the floor operation. Then, the PE $e(i,j)$ of the pixel $x(i,j)$ is determined by the Eq. (2).

$$e(i,j) = \begin{cases} x(i,j), & \text{If } i = 1 \ \& \ j = 1 \\ x(i,j-1) - x(i,j), & \text{If } i = 1 \ \& \ j \ne 1 \\ x(i-1,j) - x(i,j), & \text{If } i \ne 1 \ \& \ j = 1 \\ x(i,j) - p(i,j), & \text{If } i \ne 1 \ \& \ j \ne 1 \end{cases} \tag{2}$$

Clearly, the range of the PE $e(i,j)$ is from $-255$ to $255$, which exceeds the pixel range of a gray-scale image. If PEs are directly used to construct image, it will cause the pixel-overflow problem. To address this, we use the absolute values of PEs and a sign map to record the PEs. Let $L_s(i,j)$ be the element of the sign map $L_s$ in the $i$-th row and $j$-th column. Thus, it can determined as follows.

$$L_s(i,j) = \begin{cases} 0, & \text{If } e(i,j) \ge 0 \\ 1, & \text{Otherwise} \end{cases} \tag{3}$$

Clearly, the sign $L_s(i,j)$ can indicate that $e(i,j)$ is a negative number or a non-negative number. Moreover, the absolute values of PEs are decomposed into bit-planes for vacating room. The bit-planes of the PE image can be calculated as follows.

$$e(l,i,j) = \left\lfloor \frac{|e(i,j)|}{2^{l-1}} \right\rfloor \bmod 2, \quad l = 1, 2, \dots, 8 \tag{4}$$

where $e(l,i,j)$ is the $l$-th bit of $e(i,j)$, and $mod$ is the modular operation. Therefore, the $l$-th bit-plane of the PE image can be generated by taking all $e(l,i,j)$ values. Clearly, the PE image can be exactly recovered by using the sign map and the absolute values of PEs. Note that during image recovery, the pixels in the first row and the first column are firstly restored, then the recovered pixels are used to calculate the predicted value by using the Eq. (1) and finally the predicted value and the corresponding PE are used to recover its original pixel. To reduce the cost of storing the sign map $L_s$, the arithmetic coding is exploited to conduct compression. Figure 2 is an example of the PE calculation.

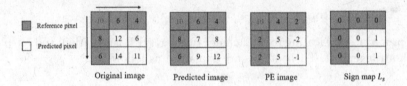

**Fig. 2.** An example of the PE calculation

**(2) Bit-Plane Block Arrangement.** As most absolute values of PEs are small values, many bits in their high bit-planes are zero. This feature can be used to conduct compression. To do so, the bit-planes are divided into non-overlapping blocks sized $k \times k$. These blocks are classified into two kinds: unique blocks (UB) and non-unique blocks (NUB). In the UB, all bits have the same value. In other words, these bits are all 1 or 0. For NUB, the values of its bits are not the same. Intuitively, the UB can be losslessly compressed by using a sign instead of record-ing all bits. In this work, a label map $L_u$ is used to mark the blocks of a bit-plane. If the block is a UB, its corresponding value in the label map is 0. Otherwise, its cor-responding value is 1. Note that the UBs can be also divided into two categories: block with bits of '0' and block with bits of '1'. To distinguish them, another sign is needed. Here, the sign of the block with bits of '0' is marked 0 and the sign of the block with bits of '1' is marked 1. In addition, the NUBs can be also divided into two categories: compressible block (CB) and non-compressible block (NCB). For a NUB, if the bit number of '0' or '1' is much bigger than that of '1' or '0', it is the CB because it can be losslessly compressed by recording the information of the few bits. Otherwise, it is the NCB. Similarly, to distinguish CB and NCB, another sign is also required. Therefore, the CB is marked with 0 and the NCB is marked with 1. In this work, a label map $L_c$ is used to mark the detailed categories of UB and NUB. Consequently, it is easy to identify the type of a block in the bit-plane by jointly using the label maps $L_u$ and $L_c$.

Classification of CB and NCB can be determined by calculating the bit cost of marking the block. If the block is the CB, some auxiliary information is required to record for exact recovery. The auxiliary information includes: Value of few bits (1 bit), Number of few bits, and Locations of few bits. Suppose that the number of few bits is $m$ in the NUB. Thus, the cost of storing the number of few bits is $\lceil \log_2 m \rceil$ bits, and the cost of storing the locations of few bits is $m \lceil \log_2 k \times k \rceil = m \lceil 2 \log_2 k \rceil$ bits, where $\lceil \cdot \rceil$ is the ceiling operation. Therefore, the total bit cost of marking the NUB $C_{\text{NUB}}$ can be determined by the following equation.

$$C_{\text{NUB}} = 1 + \lceil \log_2 m \rceil + m \lceil 2 \log_2 k \rceil \tag{5}$$

Clearly, there are $k^2$ bits in a block. Therefore, the vacated room of the NUB is $S = k^2 - C_{\text{NUB}}$ bits. If $S > 0$, the NUB is a CB. Otherwise, it is a NCB. As the fixed-length coding is used to record the number of few bits (i.e., $m$), the theoretical maximum $m$ value should be determined. For a given block size $k$, the maximum $m$ value can be determined by solving the following optimization problem.

$$m = \arg\min_{m}(S), \quad \text{s.t.} \quad S > 0 \tag{6}$$

where $m \in \{1, 2, ..., k^2/2\}$. Figure 3 presents the structure of the auxiliary information of CB. Obviously, the CB can be exactly recovered by using the auxiliary information.

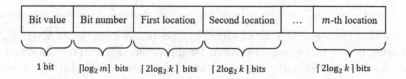

**Fig. 3.** Structure of the auxiliary information of CB

To vacate room, all blocks of a bit-plane are arranged in the below order. First, the UBs are arranged in order from left to right and top to bottom. Next, all CBs are arranged behind the UBs. Finally, all NCBs are arranged behind CBs. As the arrangement order is fixed, all blocks can be correctly placed to their original position with the guide of the label maps of $L_u$ and $L_c$ during image recovery. Since the two label maps are the auxiliary information of a bit-plane, they are compressed by arithmetic coding for reducing storage cost. Besides the label maps of $L_u$ and $L_c$, the bits of NUBs (CBs + NCBs) and the length of the bits of NUBs are both needed for correct recovery. The bits of NUBs are constructed by concatenating the auxiliary information of all CBs and the bits of all NCBs. Figure 4 presents the structure of the bits of NUBs.

| Auxiliaury information of 1st CB | Auxiliaury information of 2nd CB | ... | Auxiliaury information of $q$-th CB | Bits of 1st NCB | Bits of 2nd NCB | ... | Bits of $n$-th NCB |
|---|---|---|---|---|---|---|---|

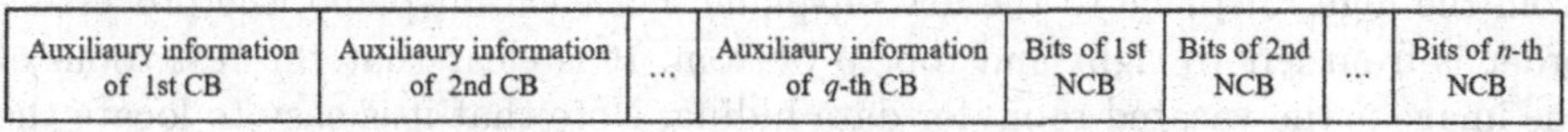

**Fig. 4.** Structure of the bits of NUBs

Clearly, the auxiliary information of a bit-plane includes: the length of the bits of NUBs (CBs + NCBs), the bits of NUBs (CBs + NCBs), the length of $L_u$, the encoded $L_u$, the length of $L_c$ and the encoded $L_c$. Here, we use $\lceil \log_2 MN \rceil$, $\lceil \log_2(\lfloor \frac{M}{k} \rfloor \lfloor \frac{N}{k} \rfloor) \rceil$ and $\lceil \log_2(\lfloor \frac{M}{k} \rfloor \lfloor \frac{N}{k} \rfloor) \rceil$ bits to record the length of the bits of NUBs, the length of $L_u$, and the length of $L_c$, respectively. Suppose that the number of the bits of NUBs is $c_1$, the bit number of the encoded $L_u$ is $c_2$ and the bit number of the encoded $L_c$ is $c_3$. Figure 5 presents the structure of the auxiliary information of a bit-plane. Therefore, the bit cost of the auxiliary information of a bit-plane can be determined by the below equation.

$$C_{\text{bitplane}} = \lceil \log_2 MN \rceil + 2 \left\lceil \log_2(\left\lfloor \frac{M}{k} \right\rfloor \left\lfloor \frac{N}{k} \right\rfloor) \right\rceil + c_1 + c_2 + c_3 \tag{7}$$

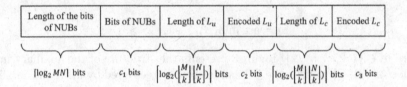

**Fig. 5.** Structure of the auxiliary information of a bit-plane

Note that there are $MN$ bits in total in a bit-plane. Let $P = MN - C_{\text{bitplane}}$. Obviously, if $P > 0$, there is still room for data embedding. Therefore, the bit-plane can be used for data hiding. Otherwise, it isn't used. In regard to the bit-plane for data hiding, it can be recovered by its auxiliary information. In regard to the unused bit-plane, its all bits should be saved. In this work, we use 1 bit to mark whether or not a bit-plane is used for data hiding, where '0' stands for use and '1' stands for unuse. Therefore, the length of the bit-plane label is 8 bits. Different from some previous RDHEI methods, the auxiliary information is not separately stored in its bit-plane. Here, all auxiliary information are concatenated to form the total auxiliary information of the image. Figure 6 presents the structure of the auxiliary information of the image. In this structure, the first part is the bit length of the total auxiliary information, which is represented by $\lceil \log_2 8MN \rceil$ bits. The second part is the bit-plane label with 8 bits. The third part is the bit length of the encoded $L_s$, which is represented by $\lceil \log_2 MN \rceil$ bits. The fourth part is the encoded $L_s$. The 5th~12th parts are the information of the 8th~1st bit-planes, respectively. Note that if the bit-plane is used for data hiding, the information is its auxiliary information. Otherwise, the information is its all bits.

After the total auxiliary information is obtained, we use it to fill the image from the high bit-plane to the low bit-plane. For each bit-plane, the processing order is from left to right and top to bottom. It is clear that the rest room of the image is the vacated room for data hiding. Note that it is easy to locate the start position of the vacated room by using the length of auxiliary information.

| Total length | Bit-plane label | Length of encoded $L_s$ | Encoded $L_s$ | Information of 8th bit-plane | Information of 7th bit-plane | ... | Information of 1st bit-plane |
|---|---|---|---|---|---|---|---|

**Fig. 6.** Structure of auxiliary information of the image

For better understanding the process of bit-plane block arrangement, an example is presented in Fig. 7, where the block size is $4 \times 4$. Figure 7(a) is a bit-plane sized $12 \times 12$, where the green blocks are UBs and the white blocks are NUBs. Figure 7 (b) is the label map $L_u$ for marking these blocks. After arranging UBs in the bit-plane, Fig. 7 (c) is available. Next, the label map $L_c$ is generated to record the arranged bit-plane (i.e., (c)), as shown in (d). After arranging CBs in the bit-plane, Fig. 7 (e) is available. By encoding the CBs, the vacated room of the bit-plane is obtained, as shown in (f). Figure (g) is the final result by filling the bits of NUBs from left to right and top to bottom.

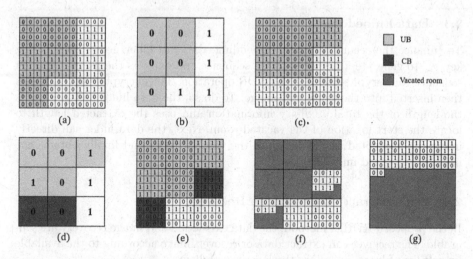

**Fig. 7.** An example of bit-plane block arrangement. (a) A bit-plane, (b) Label map $L_u$, (c) UB arrangement, (d) Label map $L_c$, (e) CB arrangement, (f) Vacated room, (g) Final result

## 2.2  Image Encryption

After the bit-plane block arrangement, the pixel $I_p(i,j)$ of the pre-processed image $I_p$ in the $(i,j)$ position can be calculated by the below equation.

$$I_p(i,j) = \sum_{l=1}^{8} I_p(l,i,j) \times 2^{l-1} \qquad (8)$$

where $I_p(l,i,j)$ is the $l$-th bit of the pixel $I_p(i,j)$ ($l = 1,2,\ldots,8$). To conduct image encryption, a pseudo-random matrix $H$ of size $M \times N$ is generated by a stream cipher via the encryption key $K_e$. Then, the bit-wise encryption is done by the below equation.

$$I_e(l,i,j) = H(l,i,j) \oplus I_p(l,i,j) \qquad (9)$$

where $\oplus$ denotes the XOR operation, $H(l,i,j)$ is the $l$-th bit of the element of $H$ in the $(i,j)$ position and $I_e(l,i,j)$ is the encrypted version of $H(l,i,j)$ ($l = 1,2,\ldots,8$). Next, the encrypted pixel can be determined by the following equation.

$$I_e(i,j) = \sum_{l=1}^{8} I_e(l,i,j) \times 2^{l-1} \qquad (10)$$

Note that the first $\lceil \log_2 8MN \rceil$ bits counted from left to right and top to bottom in the highest bit-plane are not encrypted because these bits are used to locate the start position of data hiding.

### 2.3  Data Embedding

To enhance the security of data embedding, the data hider uses a data hiding key $K_d$ to generate a pseudo-random sequence and exploits the pseudo-random sequence to encrypt secret data by XOR operation. The encrypted secret data is then inserted into the encrypted image. To do so, the data hider firstly extracts the length of the total auxiliary information and uses the extracted length to locate the start location of the vacated room. Next, the data hider can directly insert the encrypted secret data into the vacated room and finally obtains the marked encrypted image $I_{\mathrm{mark}}$.

### 2.4  Data Extraction and Image Recovery

In the proposed RDHEI method, the data extraction and image recovery are separable. The receiver can extract data or recover image according to the available knowledge of keys. There are three cases as follows:

**Case 1:** The data hiding key $K_d$ is known. In this case, the receiver can only extract the secret data, but cannot restore image. Firstly, the receiver divides the marked encrypted image into 8 bit-planes and extracts the length of the auxiliary information. According to the length of the auxiliary information, the start position of embedded secret data is then determined. Finally, the receiver can extract the encrypted secret data and find the original secret data by decrypting the encrypted secret data using the data hiding key $K_d$.

**Case 2:** The encryption key $K_e$ is known. In this case, the receiver can only restore the original image, but cannot extract the secret data. Firstly, the receiver divides the marked encrypted image into 8 bit-planes, extracts the length of the auxiliary information, and decrypts the marked image using the encryption key $K_e$. According to the total length of the auxiliary information, all auxiliary information is then obtained. After that, the receiver can restore the original image with the guide of the label maps $L_u$, $L_c$ and the sign map $L_s$.

**Case 3:** Both the data hiding key $K_d$ and the encryption key $K_e$ are known. In this case, the receiver can not only extract the secret data, but also recover the original image. The detailed operations are described in the above two cases.

## 3  Experimental Results

In this section, we firstly discuss security of the proposed RDHEI method, then analyze the embedding performance, and finally compare the proposed RDHEI method with four state-of-the-art RDHEI methods [9,11,15,16]. In the experiments, the parameter value of the block size $k$ is 4, and six benchmark gray images of size $512 \times 512$ are selected as test images, as shown in Fig. 8. Moreover, three open image databases are also exploited to further test the performance. The three databases are BOSSbase [17], BOWS-2 [18] and UCID [19]. The detailed results are described as follows.

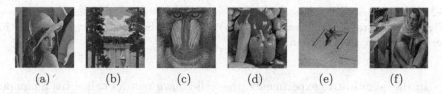

(a)        (b)        (c)        (d)        (e)        (f)

**Fig. 8.** Six benchmark gray images. (a) Lena, (b) Lake, (c) Baboon, (d) Peppers, (e) Airplane, (f) Barbara

### 3.1 Security Analysis

To validate the security of the proposed method, visual results and quantitative results are both calculated. In the visual experiments, six benchmark images are used and the results indicate that their encrypted images and the marked encrypted images are noise-like. For space limitation, the results of the Baboon are presented in the Fig. 9, where (a) is the original Baboon, (b) is the encrypted image, (c) is the marked encrypted image and (d) is the recovered Baboon. It can be seen that (b) and (c) are noisy images and there is no cue of (a). Also, the recovered Baboon is exactly the same with the original Baboon. Moreover, histograms are also calculated, as shown in Fig. 10, where (a) is the histogram of original Baboon, (b) is the histogram of the encrypted Baboon, and (c) is the histogram of the marked encrypted Baboon. It can be observed that (b) and (c) look like uniform distribution and both of them are different from (a). This also verifies that it is impossible to observe any information of the original Baboon from the encrypted Baboon and the marked encrypted Baboon. Therefore, the proposed method is secure in terms of the visual analysis.

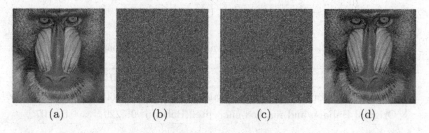

(a)              (b)              (c)              (d)

**Fig. 9.** An example of the proposed method in different stages. (a) Original Baboon, (b) Encrypted Baboon, (c) Marked encrypted Baboon, (d) Recovered Baboon

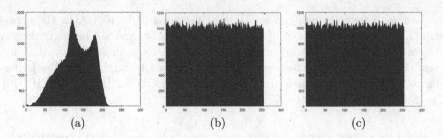

**Fig. 10.** Histograms of different images. (a) Original Baboon, (b) Encrypted Baboon, (c) Marked encrypted Baboon

In the quantitative experiments, the well-known metrics called the Shannon entropy [20], PSNR and SSIM [21] are exploited. The Shannon entropy is often used to measure image security. The range of the Shannon entropy is [0, 8] for a gray image with 256 levels. Generally, a bigger entropy means more secure image. Table 1 lists the entropies of the original Baboon, the encrypted image, and the marked Baboon image. Clearly, the entropies of the encrypted image and the marked Baboon image are close to the theoretical value 8. This illustrates that the encrypted image and the marked Baboon image generated by the proposed method are secure. In addition, the PSNR and SSIM are also computed for measuring image difference. The PSNR and SSIM are two famous metrics of image quality assessment (IQA). The maximum SSIM value is 1. For IQA task, a bigger PSNR/SSIM value means better quality of the evaluated image. On the contrary, a low PSNR/SSIM value illustrate that there is a great difference between the two images. Table 2 presents the PSNR and SSIM results. It can be seen that these PSNR and SSIM values are very small. This also demonstrates good security of the proposed method.

**Table 1.** Entropies of different images

| Image | Entropy |
|---|---|
| Original Baboon | 7.3585 |
| Encrypted Baboon | 7.9992 |
| Marked encrypted Baboon | 7.9993 |

**Table 2.** PSNR and SSIM values of different image pairs

| Image pair | PSNR (dB) | SSIM |
|---|---|---|
| Original Baboon and encrypted Baboon | 9.5260 | 0.0105 |
| Original Baboon and marked encrypted Baboon | 9.5229 | 0.0107 |

## 3.2   Embedding Performance

In this section, the embedding capacity (EC) and the embedding rate (ER) [21] are taken as the metrics for testing the embedding performance of the proposed method. Note that the EC and the ER mentioned in the experiments are the pure EC and the pure ER, respectively. Firstly, the ECs and ERs of six test images are calculated and the results are as shown in Table 3. It can be seen that the EC and ER of the proposed method are closely related with the content of the image. The smooth image will have a large EC and ER and the textural image will have a relative small EC and ER. For example, the Airplane is a smooth image and its EC and ER achieve 1070019 bits and 4.0818 bpp (bit per pixel), respectively. The Baboon has complex textures, and its EC and ER are 448817 bits and 1.7121 bpp, respectively. From Table 3, it can be found that the proposed method makes good embedding performance on the six test images.

To view our embedding performance on the large datasets, the proposed method is applied to three open image databases, i.e., BOSSbase, BOWS-2 and UCID. In this experiment, the best ER, the worst ER and the average ER are all calculated. The results are shown in Table 4. The best ERs of BOSSbase, BOWS-2 and UCID are 7.8334, 7.2348 and 5.9313 bpp, respectively. The worst ERs of BOSSbase, BOWS-2 and UCID are 0.2955, 0.9792 and 0.6853 bpp, respectively. And the average ERs of BOSSbase, BOWS-2 and UCID are 3.9773, 3.8800 and 3.1527 bpp, respectively. Our average ERs on three datasets are all bigger than 3.0 bpp. This demonstrates that the proposed method has a high embedding performance.

**Table 3.** ERs and ECs of six test images

| Image | Lena | Lake | Baboon | Peppers | Airplane | Barbara |
|-------|------|------|--------|---------|----------|---------|
| EC (bit) | 879546 | 672242 | 448817 | 789761 | 1070019 | 719192 |
| ER (bpp) | 3.3552 | 2.5644 | 1.7121 | 3.0127 | 4.0818 | 2.7435 |

**Table 4.** Our ERs on three open databases

| Database | BOSSbase | BOWS-2 | UCID |
|----------|----------|--------|------|
| Best ER | 7.8334 | 7.2348 | 5.9313 |
| Worst ER | 0.2955 | 0.9792 | 0.6853 |
| Average ER | 3.9773 | 3.8800 | 3.1527 |

## 3.3   Performance Comparison

To demonstrate advantages of the proposed method, we compare it with four popular RDHEI methods [9,11,15,16]. Here, we select these RDHEI methods for comparison because they are recently published in famous journals. In addition, the RDHEI methods [9,11] are the state-of-the-art techniques of the RRBE based RDHEI methods, and the RDHEI methods [15,16] are the state-of-the-art techniques of the VRAE based RDHEI methods. Since the proposed method and the compared RDHEI methods can losslessly restore images, we only compare the embedding performance in this section.

The ER comparison on six test images are conducted and the results are presented in Table 5, where the best results are in bold and the second best results are in italic. It can be seen that the ERs of the proposed method are bigger than those of the compared methods [11,15,16] for all test images. In addition, the ERs of the proposed method are also bigger than those of the compared method [9] for four test images, i.e., Lena, Lake, Baboon and BarBara. For Peppers and Airplane, the compared method [9] makes the best ERs and the proposed method reaches the second best ERs.

**Table 5.** ER comparison on six test images (unit:bpp)

| Method | Lena | Lake | Baboon | Peppers | Airplane | Barbara |
|---|---|---|---|---|---|---|
| Wu et al. [9] | *3.2967* | 1.8648 | 0.9660 | **3.4675** | **4.7470** | *2.6049* |
| Yin et al. [11] | 2.8699 | *2.1322* | *1.3209* | 2.5117 | 3.5136 | 2.3297 |
| Yu et al. [15] | 2.5215 | 1.8434 | 0.9695 | 2.2504 | 2.8975 | 1.9383 |
| Gao et al. [16] | 0.9812 | 0.9730 | 0.8357 | 0.9734 | 0.9961 | 0.9326 |
| Proposed | **3.3552** | **2.5644** | **1.7121** | *3.0127* | *4.0818* | **2.7435** |

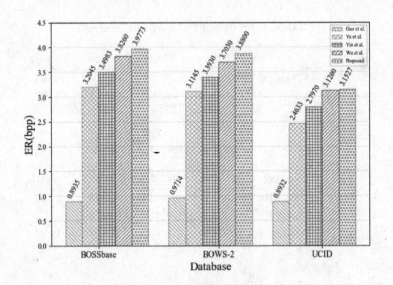

**Fig. 11.** Average ER comparison

To verify the generalized embedding performance, the average ERs of these evaluated RDHEI methods on three databases are calculated. The results are as shown in Fig. 11. The average ER of the proposed method on BOSSbase is 3.9773 bpp, and the average ERs of RDHEI methods [9,11,15,16] on BOSSbase are 3.8260, 3.4983, 3.2045, 0.8935 bpp, respectively. Our average ER on BOWS-2 is 3.8800 bpp, and the average ERs of RDHEI methods [9,11,15,16] on BOWS-2 are 3.7030, 3.3930, 3.1145, 0.9714 bpp, respectively. Our average ER on UCID is 3.1527 bpp, and the average ERs of RDHEI methods [9,11,15,16] on UCID are 3.1260, 2.7970, 2.4633, 0.8932 bpp, respectively. Obviously, the average ERs of the proposed method on the three databases are all bigger than those of the compared RDHEI methods [9,11,15,16]. These results experimentally prove that the proposed method outperforms the compared RDHEI methods [9,11,15,16] in embedding performance.

# 4   Conclusions

We have proposed a novel RDHEI method via arranging blocks of bit-planes, which can reach a high embedding performance. The proposed RDHEI method calculates the PE image, divides it into non-overlapping blocks, classifies these blocks into UBs and NUBs, and further divides NUBs into CBs and NCBs. Since the proposed RDHEI method not only uses UBs but also exploits CBs to vacate room, its embedding rate is improved. Experimental results have shown that the average ERs of the proposed RDHEI method on the BOSSbase, BOWS-2 and UCID are 3.9773, 3.8800 and 3.1527 bpp, respectively. Comparison has illustrated that the proposed RDHEI method outperforms some state-of-the-art RDHEI methods in terms of embedding rate.

**Acknowledgments.** This work is partially supported by the Guangxi Natural Science Foundation (2022GXNSFAA035506, 2021GXNSFBA196058), the National Natural Science Foundation of China (62272111, 62162006, 62062013), Guangxi "Bagui Scholar" Team for Innovation and Research, Guangxi Talent Highland Project of Big Data Intelligence and Application, and Guangxi Collaborative Innovation Center of Multi-source Information Integration and Intelligent Processing.

# References

1. Yu, C., Zhang, X.Q., Tang, Z., Xie, X.: Separable and error-free reversible data hiding in encrypted image based on two-layer pixel errors. IEEE Access **6**, 76956–76969 (2018)
2. Peng, F., Lin, Z.X., Zhang, X., Long, M.: Reversible data hiding in encrypted 2d vector graphics based on reversible mapping model for real numbers. IEEE Trans. Inf. Forensics Secur. **14**(9), 2400–2411 (2019)
3. Yao, H., Qin, C., Tang, Z., Tian, Y.: Improved dual-image reversible data hiding method using the selection strategy of shiftable pixels' coordinates with minimum distortion. Signal Process. **135**, 26–35 (2017)
4. Tang, Z., Lu, Q., Lao, H., Yu, C., Zhang, X.Q.: Error-free reversible data hiding with high capacity in encrypted image. Optik **157**, 750–760 (2018)

5. Yang, Y., Cai, X., Xiao, X., Ye, J., Shi, W.: A novel reversible data hiding with skin tone smoothing effect for face images. In: Wang, H., Zhao, X., Shi, Y., Kim, H.J., Piva, A. (eds.) IWDW 2019. LNCS, vol. 12022, pp. 204–212. Springer, Cham (2020). https://doi.org/10.1007/978-3-030-43575-2_17

6. Chen, F., Yuan, Y., Chen, Y., He, H., Qu, L.: Reversible data hiding scheme in encrypted-image based on prediction and compression coding. In: Yoo, C.D., Shi, Y.-Q., Kim, H.J., Piva, A., Kim, G. (eds.) IWDW 2018. LNCS, vol. 11378, pp. 216–229. Springer, Cham (2019). https://doi.org/10.1007/978-3-030-11389-6_17

7. Ma, K., Zhang, W., Zhao, X., Yu, N., Li, F.: Reversible data hiding in encrypted images by reserving room before encryption. IEEE Trans. Inf. Forensics Secur. **8**(3), 553–562 (2013)

8. Zhang, W., Ma, K., Yu, N.: Reversibility improved data hiding in encrypted images. Signal Process. **94**, 118–127 (2014)

9. Wu, X., Qiao, T., Xu, M., Zheng, N.: Secure reversible data hiding in encrypted images based on adaptive prediction-error labeling. Signal Process. **188**, 108200 (2021)

10. Ren, H., Lu, W., Chen, B.: Reversible data hiding in encrypted binary images by pixel prediction. Signal Process. **165**, 268–277 (2019)

11. Yin, Z., She, X., Tang, J., Luo, B.: Reversible data hiding in encrypted images based on pixel prediction and multi-MSB planes rearrangement. Signal Process. **187**, 108146 (2021)

12. Zhang, X.P.: Reversible data hiding in encrypted image. IEEE Signal Process. Lett. **18**(4), 255–258 (2011)

13. Tang, Z., Xu, S., Yao, H., Qin, C., Zhang, X.: Reversible data hiding with differential compression in encrypted image. Multimedia Tools Appl. **78**(8), 9691–9715 (2018). https://doi.org/10.1007/s11042-018-6567-3

14. Tang, Z., Pang, M., Yu, C., Fan, G., Zhang, X.Q.: Reversible data hiding for encrypted image based on adaptive prediction error coding. IET Image Proc. **15**(11), 2643–2655 (2021)

15. Yu, C., Zhang, X.Q., Li, G., Zhan, S., Tang, Z.: Reversible data hiding with adaptive difference recovery for encrypted images. Inf. Sci. **584**, 89–110 (2022)

16. Gao, G., Tong, S., Xia, Z., Shi, Y.: A universal reversible data hiding method in encrypted image based on MSB prediction and error embedding. IEEE Trans. Cloud Comput. (2022)

17. Bas, P., Filler, T., Pevný, T.: Break our steganographic system: the ins and outs of organizing boss. In: Filler, T., Pevný, T., Craver, S., Ker, A. (eds.) IH 2011. LNCS, vol. 6958, pp. 59–70. Springer, Heidelberg (2011). https://doi.org/10.1007/978-3-642-24178-9_5

18. Bas, P., Furon, T.: Image database of bows-2. Accessed 20 June 2017

19. Schaefer, G., Stich, M.: Ucid: an uncompressed color image database. In: Storage and Retrieval Methods and Applications for Multimedia 2004, vol. 5307, pp. 472–480. International Society for Optics and Photonics (2003)

20. Tang, Z., Wang, F., Zhang, X.: Image encryption based on random projection partition and chaotic system. Multimedia Tools Appl. **76**(6), 8257–8283 (2016). https://doi.org/10.1007/s11042-016-3476-1

21. Tang, Z., Xu, S., Ye, D., Wang, J., Zhang, X., Yu, C.: Real-time reversible data hiding with shifting block histogram of pixel differences in encrypted image. J. Real-Time Image Process. **16**(3), 709–724 (2018). https://doi.org/10.1007/s11554-018-0838-0

# High Capacity Reversible Data Hiding for Encrypted 3D Mesh Models Based on Topology

Yun Tang[1], Lulu Cheng[1], Wanli Lyu[1], and Zhaoxia Yin[2,3](✉)

[1] Anhui Province Key Laboratory of Multimodal Cognitive Computation,
School of Computer Science and Technology, Anhui University, Hefei 230601,
People's Republic of China
[2] Shanghai Key Laboratory of Multidimensional Information Processing, East China
Normal University, Shanghai 200241, China
[3] School of Communication and Electronic Engineering, East China Normal
University, Shanghai 200241, China
zxyin@cee.ecnu.edu.cn

**Abstract.** Reversible data hiding in encrypted domain(RDH-ED) can not only protect the privacy of 3D mesh models and embed additional data, but also recover original models and extract additional data losslessly. However, due to the insufficient use of model topology, the existing methods have not achieved satisfactory results in terms of embedding capacity. To further improve the capacity, a RDH-ED method is proposed based on the topology of the 3D mesh models, which divides the vertices into two parts: embedding set and prediction set. And after integer mapping, the embedding ability of the embedding set is calculated by the prediction set. It is then passed to the data hider for embedding additional data. Finally, the additional data and the original models can be extracted and recovered respectively by the receiver with the correct keys. Experiments declare that compared with the existing methods, this method can obtain the highest embedding capacity.

**Keywords:** Vertices division · 3D mesh model · Reversible data hiding · Encrypted models

## 1  Introduction

Data hiding [1] is a method to achieve copyright protection and covert communication by embedding additional data into the cover media and extracting the data without errors. However, the original media can not be restored losslessly by the traditional data hiding methods. It is a fatal problem in the military, medical and other fields where data integrity is strictly required. To solve this problem, reversible data hiding(RDH) is proposed.

RDH can embed additional data in the cover media and restore both media and data without errors. Over the years, researchers have devised many methods. They can be divided into three main categories: lossless compression [3], difference expansion [11] and histogram shifting [7]. In lossless compression methods,

X. Zhao et al. (Eds.): IWDW 2022, LNCS 13825, pp. 205–218, 2023.
https://doi.org/10.1007/978-3-031-25115-3_14

some features of the original media are discovered and losslessly compressed, making room for storing additional data. Difference expansion explores redundancy between adjacent pixels and embeds additional data into their difference. Histogram shifting embeds additional data by shifting peak points toward zero points in the pixel distribution histogram.

With the boom in cloud computing and privacy protection, many multimedia are uploaded, such as digital images, audio, 3D mesh models, etc., to the cloud for storage and transmission. In order to protect the privacy and security of data, media will be encrypted before uploading. In this scenario, RDH-ED has caught the attention of many researchers. There are two main categories of RDH-ED: vacating room after encryption (VRAE) [15,18,19] and reserving room before encryption (RRBE) [6,8,16]. Zhang et al. [18] introduced a RDH scheme for encrypted images based on VRAE. The encrypted images were divided into non-overlapping blocks, and each block can store 1 bit of data by flipping the 3 least significant bits (LSB) of the encrypted data. In [19], a separable RDH-ED method was proposed. In this method, a specific matrix multiplication was designed to losslessly compress the LSB of the encrypted image, which implemented the separability of data extraction and image restoration. However, the embedding ability of these methods is low due to encrypted image entropy maximization. To increase the embedding capacity and reduce the bit error rate, Ma et al. [6] first proposed a method based on RRBE, which implemented both data extraction and image recovery losslessly based on histogram shifting. Although many RDH-ED methods have obtained considerable outcomes, there is not enough research on 3D mesh models.

As an emerging digital media following images, audio and videos, 3D mesh models are applied in many fields, such as medical treatment, construction, video games, etc. To protect the privacy security and verification copyright of 3D mesh models in massive application scenarios, it is essential to study the RDH-ED methods of 3D mesh models. Jiang et al. [4] first proposed the RDH method based on VRAE for 3D mesh models in encrypted domain. However, the data extracted may be errors and the capacity of this method is not satisfactory. Shah et al. [9] and van Rensburg et al. [13] applied homomorphic encryption to RDH-ED method and embedded data in dual domain. Although homomorphic encryption increases the capacity of the method, the computational cost is high. Tsai [12] proposed a native method by spatial subdivision and space encoding. However, If the parameters are not chosen properly, errors may occur during extraction. Xu et al. [14] proposed the most significant bit(MSB) prediction, which implemented extraction and recovery separably. In [17], the embeddable room is extended to multi-MSB. However, the vertices are not fully utilized. Lyu et al. [5] made the use of vertices more fully by adding labels. However, the topology of 3D mesh models is not utilized and the division of vertices is not sufficient. Therefore, this study proposes a vertex division method based on the model topology and manages to the highest embedding capacity compared with state-of-the-art methods.

The main contributions of the proposed method are as follows:

1) We take full advantage of the topology of the 3D mesh models and propose a more reasonable vertex division method.

2) Our division method enables the correlation between vertices to be further exploited. Experiments declare that the proposed method has the highest embedding capacity compared to other methods.

The remainder of this paper is organized as follows: The related works are presented in Sect. 2. Section 3 presents the proposed method. The experimental results and analysis are shown in Sect. 4. Finally, the paper is concluded in Sect. 5.

## 2   Related Works

In this section, we introduce the types of research models and the existing RDH-ED methods for 3D mesh models.

In computer-aided design, 3D mesh models are usually represented as triangular or polygonal models, which are composed of vertex data and face data. This paper studies the most popular triangular 3D mesh models. Vertex data includes vertex coordinates and is represented as $V = \{v_i | 1 \leq i \leq n\}$ where $n$ donates the number of vertices. In the rectangular coordinate system of space, each vertex has three values in the $x$, $y$, and $z$ directions. Topological relationships between vertices are contained in face data, which is represented as $F = \{f_1, f_2, ..., f_m\}$ and $m$ is the number of the face. Each face $f$ contains three vertices. The corresponding file format is shown in Table 1.

**Table 1.** File format of a triangular mesh model.

| Vertex data | | | | Face data | |
|---|---|---|---|---|---|
| Index of vertex | $x$-axis | $y$-axis | $z$-axis | Index of face | Elements of each face |
| $v_1$ | $v_{1,x}$ | $v_{1,y}$ | $v_{1,z}$ | $f_1$ | $(v_{15}, v_2, v_1)$ |
| $v_2$ | $v_{2,x}$ | $v_{2,y}$ | $v_{2,z}$ | $f_2$ | $(v_4, v_6, v_1)$ |
| ... | ... | ... | ... | ... | ... |
| $v_{31}$ | $v_{31,x}$ | $v_{31,y}$ | $v_{31,z}$ | $f_{31}$ | $(v_6, v_1, v_3)$ |
| ... | ... | ... | ... | ... | ... |

Jiang et al. [4] introduced the RDH method to 3D mesh models in encrypted domain for the first time. The vertices were first mapped to integers and encrypted by stream cipher. After encryption, additional data was embedded into the multi-LSB. However, this method is not satisfactory in terms of embedding capacity and there may be errors in data extraction. Shah et al. [9] proposed a two-tier RDH-ED framework. The authentication information was embedded by the data sender in the first tier and additional data was embedded by the cloud server in the second tier. This method improves the embedding capacity by applying homomorphic encryption, but this method brings the file size increments and the encryption process is expensive. Van Rensburg et al. [13]

applied Paillier encryption to 3D mesh models and proposed an improved two-layer RDH method. This method minimizes file size increments as much as possible and improves the capacity. However, the original mesh models cannot be restored losslessly. Tsai [12] proposed a separable RDH method based on spatial subdivision and space encoding for encrypted 3D mesh models. However, this method has the problem of bit error during extracting data. Xu et al. [14] used the MSB prediction and proposed a new method based on RRBE, which increased the embedding capacity. Vertices were divided into embedding set and reference set by traversing the face data. The reference set is used for predicting the embedding set. After division and MSB prediction, the additional data was embedded in the MSB plane of the embedding set. Based on [14], Yin et al. [17] implemented higher embedding capacity by utilizing multi-MSB to hide data. However, this method does not adaptively assign payload to the vertices of the embedding set. Lyu et al. [5] introduced vertex labels as auxiliary information and improved the utilization of vertices. However, this method has shortcomings in the division of vertices. This method only divided the vertices by the index of vertex data and ignored the face data. To divide the vertices more reasonably, we propose a vertex division method based on model topology, which has much improved the embedding capacity compared with state-of-the-art methods.

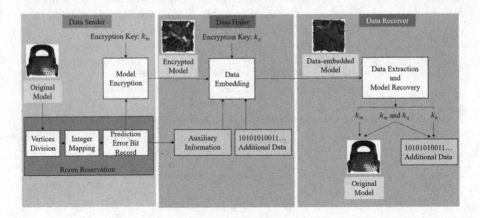

**Fig. 1.** Method flowchart.

## 3   Proposed Method

In this section, the proposed method is described in detail. As shown in Fig. 1, the proposed method consists of four parts: (1) The mesh models are preprocessed to make room for embedding additional data. (2) The 3D mesh models are encrypted by stream ciphers. (3) The encrypted additional data is embedded in the encrypted model by the data hider. (4) The additional data and original models can be extracted and recovered losslessly by the data receiver who has the corresponding key.

## 3.1   Room Reservation

**Vertices Division.** We propose a new division method and increase the number of the embedding vertices reasonably. In [4,14,16], the vertices were divided according to the face data. Each face was traversed in numbering order, and at most 1 vertex in each face was added to the embedding set. This method is not sufficient for the utilization of vertex data, and can only reach 33% of the vertex utilization for embedding. In [5], the vertices were divided according to the parity of the vertex index. This method applied more vertices to embedding additional data and the vertex utilization ratio reached 50%. However, only the vertex data was used to divide vertices. To divide the vertices more reasonably, a native division method combining vertex data and face data is proposed.

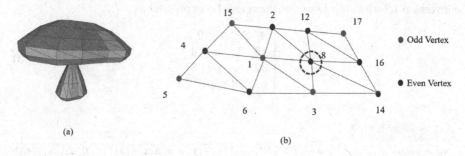

(a)

(b)

**Fig. 2.** (a) *Mushroom*, (b) Labeled local topology of *Mushroom*.

In our method, the vertices are divided into two parts first according to the parity of the vertex index. For the convenience of introduction, here we select the even-numbered part as the prediction set $S_p$ and the odd-numbered part as the embedding set $S_e$. Some vertices in $S_p$ are redundant and unsuitable to use as prediction vertices. We traverse the vertices in $S_p$ one by one and filter out the vertices that are suitable for embedding additional data. In order to ensure that some vertices in $S_p$ with excellent predictive ability are not classified into the $S_e$, vertices are added to the $S_e$ only when the number of odd-numbered vertices around the vertices does not exceed twice the number of even-numbered vertices. Besides, in order to ensure the prediction accuracy, the vertices must also be guaranteed that they can be predicted by at least 2 vertices. An example of the vertices screening process is shown in Fig. 2. The vertex $v_8$ has four even neighbor vertices which means that this vertex can be accurately predicted with multi-MSB and the number of odd-neighbored vertices does not exceed twice the number of even-numbered vertices. Therefore, $v_8$ is more suitable for the $S_e$. After that, the predicted vertices for $v_1$ will have $v_8$ removed.

**Integer Mapping.** Uncompressed vertex coordinate values are stored in the computer as 32-bit floating point numbers. [2] suggested that 3D mesh models do not require this level of accuracy for most applications. Therefore, lossy compression is performed on the values of vertices. The value of vertices is expanded to between $-10^p$ and $10^p$, where $p\epsilon(1, 33)$. The expanded value is processed into an integer by

$$v'_{i,j} = \lfloor v_{i,j} \times 10^p \rfloor, \ i = 1, 2, ..., n, \ j\epsilon(x, y, z), \tag{1}$$

where $\lfloor . \rfloor$ is the round down function. The receiver can restore the coordinates to floating point coordinates by

$$v''_{i,j} = v'_{i,j}/10^p, \ i = 1, 2, ..., n, \ j\epsilon(x, y, z). \tag{2}$$

The bit length $l$ of the integer is determined by the value of $p$, and the conversion relationship between them can be expressed as

$$l = \begin{cases} 8, \ 1 \le p \le 2 \\ 16, \ 3 \le p \le 4 \\ 32, \ 5 \le p \le 8 \\ 64, \ 10 \le p \le 33. \end{cases} \tag{3}$$

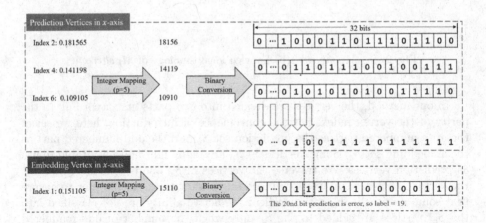

**Fig. 3.** Prediction error labels for $v_1$ in $x$-axis.

**Prediction Error Bit Record.** The embedding vertex is predicted by the vertices around it. In Fig. 3, the process of predicting error labels is illustrated in detail. The $v_1$ has three neighbor vertices in $S_p$, which are $v_2, v_4$ and $v_6$. From the MSB to the LSB, the predicted value of each bit plane is determined according to the mode of all corresponding bit planes of the prediction vertices. If the number of 0 and 1 is the same, 1 is chosen as the final value. For example, in the 17th digit of index 2, 4 and 6, 0 appears the most. So the predicted value is chosen to be 0. The embeddable length is obtained by comparing the predicted value with $v_1$ and records as a label. Since a vertex has coordinates in three directions, the label selection for a vertex will finally be the minimum of the three directions. In Fig. 3, $v_1$ has a label value of 19 in $x$-axis. But the final label value of $v_1$ still needs to be determined by the other two directions $y$-axis and $z$-axis. The embeddable length of each vertex is recorded in the location map, which is compressed by arithmetic coding and then embedded into the 3D mesh models as auxiliary information.

## 3.2   Model Encryption

The 3D mesh model is encrypted by stream cipher in this paper. First, the coordinate of each vertex is converted to a binary representation. Each binary bit can be obtained by

$$b_{i,j,k} = \lfloor v'_{i,j}/2^k \rfloor \bmod 2, \ k = 0, 1, ..., l - 1. \tag{4}$$

Then, the data sender uses the stream cipher function to generate pseudo-random bits $c_{i,j,k}$, and the encrypted bits are obtained by

$$e_{i,j,k} = b_{i,j,k} \oplus c_{i,j,k}, \ k = 0, 1, ..., l - 1, \tag{5}$$

where $\oplus$ means exclusive OR. Finally, the encrypted vertices can get by

$$v'''_{i,j,k} = \sum_{k=0}^{l-1} e_{i,j,k} \times 10^k. \tag{6}$$

## 3.3   Data Embedding

The process of data embedding is shown in Fig. 4. The room obtained in Sect. 3.1 is used for data hiding. First, the data hider divides the binary representation of vertex values in $S_e$ into two parts according to the location map, the embeddable part and the non-embeddable part. Then, the encrypted additional data and auxiliary information are embedded into the embeddable part by substitution. Finally, the vertices are reconstructed.

## 3.4   Data Extraction and Model Recovery

The data receiver who has the correct key can decrypt the encrypted data. The additional data can be extracted with the key $k_a$ and original mesh models can be recovered with the key $k_m$. Therefore, here are three possible scenarios:

Case 1: The receiver only holds $k_a$. All the additional data can be extracted from the encrypted model and plaintext data is decrypted with $k_a$.

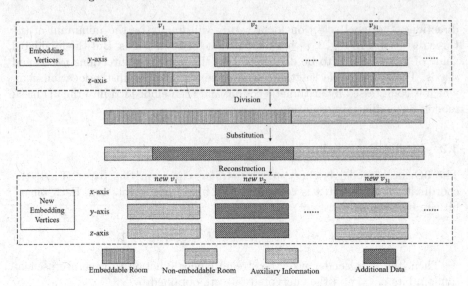

**Fig. 4.** Process of data embedding.

Case 2: The receiver only holds $k_m$. The data receiver extracts the auxiliary data and decrypts the mesh models through the key $k_m$. However, the embedding set in the decrypted mesh model is not the same as the original one. Then, the prediction set is applied to restore the embedding set with the help of the auxiliary data.

Case 3: The receiver holds both keys $k_a$ and $k_m$. At this time, the data receiver can obtain additional data and restore the original models at the same time. The procedure is the same as in case 1 and case 2, but model decryption is performed after data extraction.

## 4   Experiment Results and Analysis

### 4.1   Visual Quality and Quantitative Analysis

Four test mesh models are demonstrated in Fig. 5. Despite the slight distortion introduced to the models during the preprocess stage, there is little difference between the original models and the recovered models based on human visual observations. This shows that our method is feasible in the application. To further demonstrate the feasibility of the method, we introduce signal-to-noise ratio(SNR) and Hausdorff distance to quantitatively measure the gap between the original models and the recovered ones.

SNR can calculate the gap between two models. The larger the SNR, the smaller the gap between the two models. SNR can be obtained by

$$\text{SNR} = 10\log_{10} \frac{\sum_{i=1}^{N}[(v_{i,x} - \bar{v}_x)^2 + (v_{i,y} - \bar{v}_y)^2 + (v_{i,z} - \bar{v}_z)^2]}{\sum_{i=1}^{N}[(v_{i,x}'' - v_{i,x})^2 + (v_{i,y}'' - v_{i,y})^2 + (v_{i,z}'' - v_{i,z})^2]}, \quad (7)$$

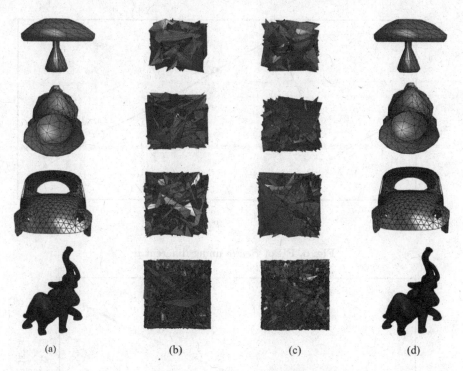

**Fig. 5.** Models presentation at different stages($p = 5$): (a) original models; (b) encrypted models; (c) data-embedded models; (d) recovered models.

where $v_{i,x}, v_{i,y}, v_{i,z}$ are the values of original coordinates, $\bar{v}_x, \bar{v}_y, \bar{v}_z$ are the average values of original coordinates, and $v''_{i,x}, v''_{i,y}, v''_{i,z}$ are the values of recovery coordinates.

Hausdorff distance measures the maximum distance value among the closest distances between two model vertices. The smaller the Hausdorff distance, the smaller the gap between the two models. Assuming there are two sets $A = (a_1, a_2, ..., a_u)$ and $B = (b_1, b_2, ..., b_u)$, Hausdorff distance can be calculated as

$$
\begin{aligned}
D(A, B) &= \max(d(A, B), d(B, A)), \\
d(A, B) &= \max_{a \in A} \min_{b \in B} \|a - b\|, \\
d(B, A) &= \max_{b \in B} \min_{a \in A} \|b - a\|,
\end{aligned}
\tag{8}
$$

where $\|.\|$ is the Euclidean distance, and $u$ denotes the number of elements in each set.

The ER of the model *Beetle* under different $p$ values is given in Fig. 6 and the values of SNR and Hausdorff distance of model *Beetle* under different p values are shown in Fig. 7. It can be seen that as p becomes larger, although the disturbance introduced into the model becomes smaller, ER also becomes smaller.

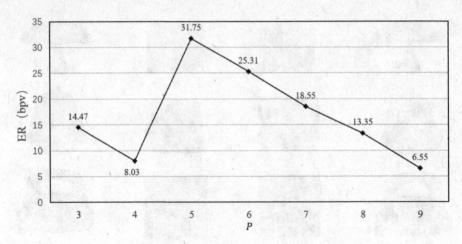

**Fig. 6.** ER of *Beetle* under different $p$

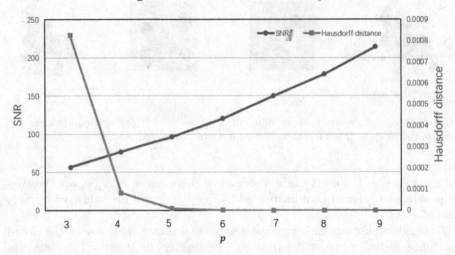

**Fig. 7.** ER of *Beetle* under different $p$

Under the trade-off of embedding capacity and the requirements of practical applications, the value of $p$ is to be 5. The results of four test models are shown in Table 2. As can be seen in the table, the SNR of the four test models has reached a high value, and the Hausdorff distance is almost zero.

**Table 2.** SNR and Hausdorff distance of four test models ($p = 5$).

| Test models | Number of vertex | Number of face | SNR | Hausdorff distance($10^{-6}$) |
|---|---|---|---|---|
| Mushroom | 226 | 448 | 102.26 | 8.12 |
| Mannequin | 428 | 839 | 130.93 | 4.06 |
| Beetle | 988 | 1763 | 96.20 | 8.66 |
| Elephant | 24955 | 49918 | 95.97 | 8.66 |

## 4.2 Capacity Analysis

The embedding rate(ER) is one of the important indicators of the RDH-ED methods for 3D mesh models in encrypted domain, which is presented by bit per vertex(bpv). ER can be calculated by

$$\mathrm{ER} = \frac{l_p - l_{ai}}{n}, \tag{9}$$

where $l_p$ is the total capacity and $l_{ai}$ is the length of auxiliary information.

**Table 3.** Comparison of ER(bpv) on four test models.

| Test models | [4] | [9] | [13] | [12] | [14] | [17] | [5] | **Ours** |
|---|---|---|---|---|---|---|---|---|
| Mushroom | 0.45 | 6 | 13 | 7.68 | 1.34 | 16.72 | 21.76 | **22.53** |
| Mannequin | 0.34 | 6 | 13 | 7.68 | 0.95 | 13.66 | 18.05 | **24.08** |
| Beetle | 0.35 | 6 | 13 | 7.68 | 0.98 | 16.51 | 23.55 | **31.75** |
| Elephants | 0.34 | 6 | 13 | 7.68 | 1.02 | 18.12 | 27.96 | **38.93** |

The comparison of ER on four test models is shown in Table 3. Jiang et al. [4] used multi-LSB for data embedding. The embedding rate of this method is not high enough, and it does not exceed 1 bpv at most. Shah et al. [9] designed a two-layer embedding method. Additional data was embedded in the first layer through histogram expansion and shifting techniques. The second layer utilized the self-blinding property of the Paillier cryptosystem to embed additional data. The final embedding rate of this method is 6 bpv. Compared with [9,13] adjusted the encryption bit length of the coordinates to reduce the file size increment. Tsai [12] exploited the ratio of vertex value normalization for model encryption and data embedding and realized 7.68 bpv. Xu et al. [14] proposed to embed data in MSB. Although the embedding rate is improved, the relationship between the adjacent vertices has not been used fully. Yin et al. [17] extended the embedding range to multi-MSB and achieved a substantial increase in capacity. Lyu et al. [5] further improved the embedding capacity and improved the utilization of vertices by labeling vertices. Due to the topology of 3D mesh models, the same vertex can be included in multiple faces and associated with multiple vertices.

Furthermore, many correct prediction bits can be obtained with the help of two or three vertices by rationally utilizing the strong multi-MSB correlation between adjacent vertices. However, among the aforementioned methods, the topological structure of the models is not used to divide the vertices and the division of vertices is not reasonable enough. The proposed method addresses this issue and implemented the state-of-the-art results on the average ER. The division results of the embedded vertices and predicted vertices on four test models are displayed in Table 4. Except for the *Mushroom* model, the number of embedded vertices does not increase greatly after vertex division due to the small number of vertices and faces, the other three models all obtained a high vertex utilization ratio. The utilization of model topology increases the number of embedded vertices, and finally the ER of the models is improved.

**Table 4.** Results of vertices division on four test models ($p = 5$).

| Test models | Number of vertex | Number of face | Number of embedding vertex | Vertex utilization ratio(%) |
|---|---|---|---|---|
| Mushroom | 226 | 448 | 121 | 53 |
| Mannequin | 428 | 839 | 316 | 73 |
| Beetle | 988 | 1763 | 705 | 71 |
| Elephant | 24955 | 49918 | 18935 | 75 |

To be more persuasive, the proposed method is tested on the Princeton dataset [10] which consists of 380 different 3D mesh models. The caparison with other methods is shown in Fig. 8 and our method also obtains state-of-the-art experimental results on the average ER.

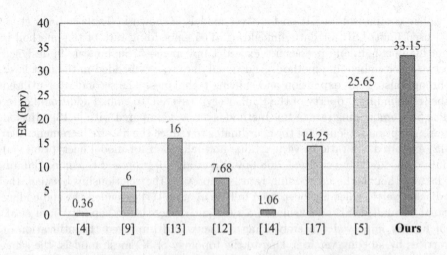

**Fig. 8.** Average capacity comparison with other methods on the Princeton dataset.

**Table 5.** Comparison of features.

| Features | Encryption method | Computation complexity | Separability | Data error |
|---|---|---|---|---|
| [4] | Stream cipher encryption | Low | No | Yes |
| [9] | Homomorphic encryption | High | No | No |
| [13] | Homomorphic encryption | High | No | No |
| [12] | Stream cipher encryption | Low | Yes | Yes |
| [14] | Stream cipher encryption | Low | Yes | No |
| [17] | Stream cipher encryption | Low | Yes | No |
| [5] | Stream cipher encryption | Low | Yes | No |
| **Ours** | **Stream cipher encryption** | **Low** | **Yes** | **No** |

## 4.3  Features Analysis

The comparison of features with other methods is shown in Table 5. Jiang et al. [4] used the smoothness function, which would lead to errors in data extraction, making the model and data inseparable during recovery and extraction. In [9, 13], it is expensive to apply homomorphic encryption to the models and the computational complexity of these methods is very high. In [12], the selection of a suitable threshold is crucial, otherwise the data extraction will be wrong. [5,14,16] are MSB-based methods that make room for embedding additional data by exploiting the correlations between vertices. The proposed method can not only recover the model and additional data reversibly, but also recover and extract them separably. Besides, the computational complexity of using stream cipher encryption is lower than that of homomorphic encryption.

## 5  Conclusion

In this paper, we study a RDH method for 3D mesh models with high capacity in encrypted domain. The vertices are divided into two parts reasonably by utilizing the vertex data and face data. One is used for data embedding, the other one is used for model recovery. The experiments demonstrate that the proposed method reaches state-of-the-art performance. In the future, we consider to enhance the robustness of the method to resist channel interference of data during network transmission. Besides, a method that does not require the transmission of auxiliary information from the data sender to the data hider is also our next research direction.

**Acknowledgements.** This research work is partly supported by National Natural Science Foundation of China (62172001, 61872003).

## References

1. Barton, J.M.: Method and apparatus for embedding authentication information within digital data. US Patent 5,646,997, 8 Jul 1997

2. Deering, M.: Geometry compression. In: Proceedings of the 22nd Annual Conference on Computer Graphics and Interactive Techniques, pp. 13–20 (1995)
3. Fridrich, J., Goljan, M., Du, R.: Lossless data embedding for all image formats. In: Security and Watermarking of Multimedia Contents IV, vol. 4675, pp. 572–583. SPIE (2002)
4. Jiang, R., Zhou, H., Zhang, W., Yu, N.: Reversible data hiding in encrypted three-dimensional mesh models. IEEE Trans. Multimedia **20**(1), 55–67 (2017)
5. Lyu, W.L., Cheng, L., Yin, Z.: High-capacity reversible data hiding in encrypted 3D mesh models based on multi-MSB prediction. Signal Process. **201**, 108686 (2022)
6. Ma, K., Zhang, W., Zhao, X., Yu, N., Li, F.: Reversible data hiding in encrypted images by reserving room before encryption. IEEE Trans. Inf. Forensics Secur. **8**(3), 553–562 (2013)
7. Ni, Z., Shi, Y.Q., Ansari, N., Su, W.: Reversible data hiding. IEEE Trans. Circuits Syst. Video Technol. **16**(3), 354–362 (2006)
8. Puteaux, P., Puech, W.: An efficient MSB prediction-based method for high-capacity reversible data hiding in encrypted images. IEEE Trans. Inf. Forensics Secur. **13**(7), 1670–1681 (2018)
9. Shah, M., Zhang, W., Hu, H., Zhou, H., Mahmood, T.: Homomorphic encryption-based reversible data hiding for 3D mesh models. Arab. J. Sci. Eng. **43**(12), 8145–8157 (2018). https://doi.org/10.1007/s13369-018-3354-4
10. Shilane, P., Min, P., Kazhdan, M., Funkhouser, T.: The Princeton shape benchmark. In: Proceedings Shape Modeling Applications, 2004, pp. 167–178. IEEE (2004)
11. Tian, J.: Reversible data embedding using a difference expansion. IEEE Trans. Circuits Syst. Video Technol. **13**(8), 890–896 (2003)
12. Tsai, Y.Y.: Separable reversible data hiding for encrypted three-dimensional models based on spatial subdivision and space encoding. IEEE Trans. Multimedia **23**, 2286–2296 (2020)
13. Van Rensburg, B.J., Puteaux, P., Puech, W., Pedeboy, J.P.: Homomorphic two tier reversible data hiding in encrypted 3D objects. In: 2021 IEEE International Conference on Image Processing (ICIP), pp. 3068–3072. IEEE (2021)
14. Xu, N., Tang, J., Luo, B., Yin, Z.: Separable reversible data hiding based on integer mapping and MSB prediction for encrypted 3D mesh models. Cogn. Comput. **14**(3), 1172–1181 (2022). https://doi.org/10.1007/s12559-021-09919-5
15. Yin, Z., Luo, B., Hong, W.: Separable and error-free reversible data hiding in encrypted image with high payload. Sci. World J. **2014** (2014)
16. Yin, Z., Xiang, Y., Zhang, X.: Reversible data hiding in encrypted images based on multi-MSB prediction and Huffman coding. IEEE Trans. Multimedia **22**(4), 874–884 (2019)
17. Yin, Z., Xu, N., Wang, F., Cheng, L., Luo, B.: Separable reversible data hiding based on integer mapping and multi-MSB prediction for encrypted 3D mesh models. In: Ma, H., et al. (eds.) PRCV 2021. LNCS, vol. 13020, pp. 336–348. Springer, Cham (2021). https://doi.org/10.1007/978-3-030-88007-1_28
18. Zhang, X.: Reversible data hiding in encrypted image. IEEE Signal Process. Lett. **18**(4), 255–258 (2011)
19. Zhang, X.: Separable reversible data hiding in encrypted image. IEEE Trans. Inf. Forensics Secur. **7**(2), 826–832 (2011)

# Author Index

Printed in the United States
by Baker & Taylor Publisher Services